Contents

Preface v

How to use this unit vi

Introduction to this unit vi

Part I Homeostasis, adaptation and evolution **1**

Section 1 Survival **1**

1.1 Introduction and objectives 1

1.2 Selection pressures 1

1.2.1 Fitness 1

1.2.2 The abiotic and biotic environments 2

1.3 Group versus individual selection 3

1.4 Short-term versus long-term pressures on survival 3

1.5 References and further reading 4

Section 2 Regulatory mechanisms (1) The control of the internal environment – homeostasis **5**

2.1 Introduction and objectives 5

2.2 Historical aspects 5

2.3 Conformers and regulators 6

2.4 Equilibrium 7

2.5 General principles of homeostasis 7

2.6 Feedback mechanisms 8

2.7 References and further reading 9

Section 3 Regulatory mechanisms (2) The response to environmental changes **10**

3.1 Introduction and objectives 10

3.2 Physiological strategies 10

3.2.1 Homeothermy and heterothermy 10

3.2.2 Hibernation 11

3.2.3 Heterothermy 11

3.2.4 Regional heterothermy 12

3.2.5 Adaptations to decreased oxygen availability at high altitude 12

3.2.6 Adaptation to decreased oxygen availability in diving mammals 13

3.2.7 Failure of regulation of body temperature – fever 14

3.3 Behavioural strategies 15

3.3.1 Homeotherm behaviour 15

3.3.2 Poikilotherm behaviour 16

3.3.3 Social organisation and survival 16

3.3.4 Migration: sur[illegible] 16

3.4 References and further reading 17

Section 4 Adaptation – evidence for evolution **18**

4.1 Introduction and objectives 18

4.2 Adaptation 18

4.2.1 Related adaptations 20

4.2.2 Niches 20

4.3 Descent with modification 21

4.4 Fossil evidence 21

4.4.1 What is a fossil? 21

4.4.2 The fossil record 22

4.4.3 *Archaeopteryx* 22

4.4.4 The horse 25

4.5 Origins, replacements and extinctions 28

4.5.1 The origin of the fossil record 28

4.5.2 Replacements and extinctions 29

4.6 Adaptive radiation 31

4.6.1 Basic principles 31

4.6.2 Adaptive radiation in mammals 32

4.7 References and further reading 35

Section 5 Evolution **36**

5.1 Introduction and objectives 36

5.2 Evolutionary theories 36

5.2.1 Lamarck 37

5.2.2 Darwin 38

5.3 Natural selection 39

5.3.1 Natural selection in action 39

5.3.2 Computer simulation of natural selection 40

5.4 Speciation 40

5.4.1 Allopatric speciation 41

5.4.2 Parapatric speciation 41

5.4.3 Polyploidy 41

5.5 Reproductive isolation mechanisms 42

5.5.1 Temporal isolation 42

5.5.2 Ecological isolation 42

5.5.3 Behavioural isolation 42

5.5.4 Gametic isolation 43

5.5.5 Mechanical isolation 43

5.5.6 Hybrid isolation 43

5.6 Extension: Galapagos finches 44

5.7 References and further reading 45

Section 6 Interactions **46**

6.1 Introduction 46

6.2 Types of interaction 46

6.3 Interactions within populations 46

6.3.1 The concepts of the individual and the colony 47

6.3.2 Intraspecific interactions between individuals 48

6.3.3 Male–female interaction 48
6.3.4 Male–male interaction 49
6.3.5 Other intraspecific interactions and pheromones 49
6.3.6 Populations and societies 50
6.3.7 Concluding remarks 51
6.4 Interactions between populations 51
6.4.1 Competition 51
6.4.2 Neutralism 51
6.4.3 Mutualism 52
6.4.4 Protocooperation and commensalism 53
6.4.5 Predation 54
6.4.6 Parasitism 56
6.4.7 Amensalism 59
6.4.8 Concluding remarks 59
6.5 Interactions between taxa 59
6.5.1 *Heliconius* butterflies and passion vines: an example of co-evolution 60
6.6 References and further reading 61

Part II Thematic review **62**

Introduction 62
List of abbreviated review tasks 64

Section 7 Levels of organisation **66**
7.1 Introduction 66
7.2 Cellular organisation 66
7.3 Multicellular organisation 67
7.3.1 Colonial level 67
7.3.2 Tissue level 68
7.3.3 Tissue systems 68
7.3.4 Organ level 68
7.3.5 Organ systems level 69
7.4 Group organisation 69
7.4.1 Species 69
7.4.2 Populations 69
7.4.3 Communities 70
7.4.4 Ecosystems 70
7.4.5 Biomes 70
7.4.6 Biosphere 70

Section 8 Chemical action **72**
8.1 Introduction 72
8.2 Chemicals of life 72
8.2.1 Inorganic compounds 72
8.2.2 Organic compounds 72
8.3 Energy liberation 73
8.3.1 Rearrangement and dehydrogenation reactions 74
8.3.2 Oxidative phosphorylation 74
8.3.3 Photophosphorylation 76
8.4 Matter availability 77
8.4.1 Introduction 77
8.4.2 Synthesis of body substances 77
8.4.3 Obtaining raw materials 79
8.5 Molecular movement 79
8.5.1 Diffusion 79
8.5.2 Active transport 80
8.5.3 Mass flow 80
8.6 Chemical action initiating activities 80
8.6.1 Receptors 81
8.6.2 Nervous communication 81
8.6.3 Effectors 81

Section 9 Equilibria **82**
9.1 Introduction 82
9.2 Molecular equilibrium 83
9.3 Constancy of internal conditions 83
9.3.1 Physical and chemical conditions 83
9.3.2 Biological integrity 84
9.4 Offspring stability 85
9.5 Population stability 85
9.5.1 Population characteristics 85
9.5.2 Population size 85
9.6 Ecosystem composition 87

Section 10 Adaptation **88**
10.1 Introduction 88
10.2 The origins of new characteristics 88
10.2.1 Reproduction of the genome 88
10.2.2 Development of characteristics 89
10.3 Variation to adaptation 89
10.3.1 Natural selection 89
10.4 Some major adaptations of living things 90

Section 11 Interaction **91**
11.1 Introduction 91
11.2 Interaction producing life 91
11.2.1 Interaction in embryos 91
11.2.2 Maturation 91
11.3 Interactions within the individual 92
11.4 Interactions between individuals 92
11.5 Interactions between living things and the environment 92

Section 12 Answers to self-assessment questions **93**

Index 99

Preface

The Inner London Education Authority's Advanced Biology Alternative Learning (ABAL) project has been developed as a response to changes which have taken place in the organisation of secondary education and the curriculum. The project is the work of a group of biology teachers seconded from ILEA secondary schools. ABAL began in 1978 and since then has undergone extensive trials in schools and colleges of further education. The materials have been produced to help teachers meet the needs of new teaching situations and provide an effective method of learning for students.

Teachers new to A-level teaching or experienced teachers involved in reorganisation of schools due to the changes in population face many problems. These include the sharing of staff and pupils between existing schools and the variety of backgrounds and abilities of pupils starting A-level courses whether at schools, sixth form centres or colleges. Many of the students will be studying a wide range of courses, which in some cases will be a mixture of science, arts and humanities.

The ABAL individualised learning materials offer a guided approach to A-level biology and can be used to form a coherent base in many teaching situations. The materials are organised so that teachers can prepare study programmes suited to their own students. The separation of core and extension work enables the academic needs of all students to be satisfied. Teachers are essential to the success of this course, not only in using their traditional skills, but for organising resources and solving individual problems. They act as personal tutors, and monitor the progress of each student as he or she proceeds through the course.

The materials aim to help the students develop and improve their personal study skills, enabling them to work more effectively and become more actively involved and responsible for their own learning and assessment. This approach allows the students to develop a sound understanding of fundamental biological concepts.

Acknowledgements

Figures: 1, from M.B.V. Roberts (1982) *Biology: a functional approach,* 3rd ed., Thomas Nelson & Sons; 3, August Krogh *Osmotic regulation in aquatic animals,* CUP; 9, by permission of F. Reed Hainsworth and the American Association for the Advancement of Science; 10, 11, 13, Knut Schmidt-Neilsen, Dept. Zoology, New Jersey; 14, M.J. Kluger, the *Journal of Physiology;* 15, M.J. Kluger, reprinted by permission from *Nature,* vol. 252, no. 5483, p.473, © 1974, Macmillan Journals Ltd; 16, Knut Schmidt-Neilsen, reprinted with permission from Vermogen van het Koninklijk Belgisch Instituut voor Natuurwetenschappen; 18, the Slide Centre; 19, John W. Kimball *Biology,* Addison-Weseley Publishing Company, USA; 20, A. Langham; 21, Cecil G. Trew (1956) *The horse through the ages,* permission from Methuen & Company; 22, Bruce J. MacFadden, Journal of Palaeobiology; 23, 55, L. Margulis (1981) *Symbiosis in cell evolution* © W.H. Freeman & Co; 24, 52, Natural History Photographic Agency; 25, British Museum (Natural History); 26, 29, National Science Curriculum Materials Project, Australia, Jacaranda Wiley Ltd; 31, P. Stanbury *Looking at mammals,* Heinemann Educational; 32, 34 (orang utan, hedgehog, fruit bat, mouse), Zoological Society of London; 34 (tiger, pig), J. Allan Cash Photolibrary; 35, J. Baker & G. Allen (1982) *The study of biology,* © Addison Weseley Publishing Co. USA; 39, H.B.D. Kettlewell, *Darwin's missing evidence, Scientific American,* March 1959; 40, with permission from P. Calow *Evolutionary principles,* Blackie & Son; 42, permission from *New Scientist,* from *What doubts evolution,* Mark Ridly, 25.6.1981; 43, 44, 45, 46, 47, 48, 49, 50, from *Origin of species,* British Museum (Natural History) & CUP; 53, Eric R. Pianka *Evolutionary Ecology* © 1983 by Harper & Row Publishers Inc. after Odum 1959 & Haskell 1947; 54, 59, 60, 63, 64*b* Marcus Barbor; 56, 57, 58, Phillip J. Whitfield (1979) *The biology of parasitism,* Edward Arnold; 61, 62, J.H. Lawton & Sir R. Southwood (1984) *Insects on plants,* Blackwell Scientific Publications.

How to use this unit

This is not a textbook. It is a guide that will help you learn as effectively as possible. As you work through it, you will be directed to practical work, audio-visual resources and other materials. There are sections of text in this guide which are to be read as any other book, but much of the guide is concerned with helping you through activities designed to produce effective learning. The following list gives details of the ways in which the unit is organised.

(1) Objectives

Objectives are stated at the beginning of each section. They are important because they tell you what you should be able to do when you have finished working through the section. They should give you extra help in organising your learning. In particular, you should check after working through each section that you can achieve all the stated objectives and that you have notes which cover them.

(2) Self-assessment questions *(SAQ)*

These are designed to help you think about what you are reading. You should always write down answers to self-assessment questions and then check them immediately with those answers given at the back of this unit. If you do not understand a question and answer, make a note of it and discuss it with your tutor at the earliest opportunity.

(3) Computer simulations

A number of activities in this unit refer to computer simulations which may be available from your tutor. They deal with topics which cannot be covered easily in text or practical work, as well as providing a change from the normal type of learning activities. This should help in motivating you.

(4) Extension work

This work is provided for several reasons: to provide additional material of general interest, to provide more detailed treatment of some topics, to provide more searching questions that will make demands on your powers of thinking and reasoning.

Study and practical skills

This unit is in two parts. Neither part includes any laboratory practical work. In part I you will continue to use the study skills that you have employed in previous units, including the use of a computer exercise. In part II a new study skill, **review**, is introduced. This skill and its purpose is described in the introduction to part II (page 62).

Pre-knowledge for this unit

You should have covered all the work of the previous nine ABAL units.

Introduction to this unit

This final unit is divided into two parts. In part I, four important concepts (homeostasis, adaptation, evolution and interactions) that have been touched upon in other units, are considered. The ideas are brought together here and related to the individual organism and the whole group in terms of short- and long-term survival.

In part II, the thematic review, five basic principles or themes are used to highlight the different areas of biology you have studied. The sections reorganise the usual topic areas to put into perspective the mass of finely detailed information you will have acquired. This biological review emphasises the unity of living things by focussing on themes basic to the continuation of life.

Part I Homeostasis, adaptation and evolution

Section 1 Survival

1.1 Introduction and objectives

This section is concerned with the challenges which are continually made upon cells, individual organisms and groups of organisms.

After completing this section you should be able to do the following.

(*a*) Define the following terms.
Artificial selection, environmental change, biotic change, selection pressure, biological fitness, variant, group selection, individual selection, altruism, kin selection, homeostasis, evolution

(*b*) Distinguish physical (abiotic) from biological (biotic) environments.

(*c*) Explain the significance of long- and short-term survival.

1.2 Selection pressures

Scientists have always been interested in the incredible variety that exists in the plant and animal kingdoms. It was from this interest that the study of genetics originated at the turn of the twentieth century. As you have progressed through this course, the question as to why there is such a variety of living organisms surviving in so many varied habitats may have crossed your mind. You will recall from your study of genetics that new breeds of plants and animals for agriculture are developed by using methods of **artificial selection**. Plant breeders and farmers may deliberately choose as parents those individuals with a desirable phenotype. Breeding is not at random, useful traits are selected to alter and improve the genotype.

In nature, a similar process occurs. **Natural selection** takes place of those traits which are favourable to survival. It is therefore interesting to see that those characteristics of living things which distinguish them from non-living matter are primarily concerned with survival.

Under natural conditions, life processes are continuously threatened by changes in environmental factors. For any individual, survival is only possible if the type and size of environmental change can be tolerated or avoided by that individual. The individuals which, through genetic selection, survive and continue to breed, are those which can withstand the pressure of change. There are examples like the modern domestic turkey where artificial selection has been encouraged to such an extent that, without the aid of the breeder, survival of the group would not continue because due to their increased size and weight, natural breeding is almost impossible.

Survival depends on the ability of organisms to adjust to changes in their abiotic and biotic environments. These changes constitute the selection pressures which constantly challenge life processes at all levels.

SAQ 1 Define the term selection pressure in your own words.

1.2.1 Fitness

It was the idea of a process of selection operating in nature that led Charles Darwin to propose his theory of evolution. Indeed, the full title of his book on this subject is *The origin of species by means of natural selection or the preservation of favoured races in the struggle for life*. In short, Darwin was not the first biologist to suggest that species were subject to change, but he was the first to propose a mechanism for evolution.

Sometimes the process of natural selection is summarised in the phrase 'survival of the fittest'.

This can be very misleading. The actual phrase 'survival of the fittest' was coined by the philosopher and sociologist Herbert Spencer. Perhaps it is unfortunate that Darwin then decided to borrow the phrase, since it has become the source of much misunderstanding. Nowadays, the term fitness, in biology, does not have its everyday meaning (that is, keeping fit, being healthy) nor does it have quite the meaning it had for Darwin or Spencer. **Fitness** is a measure of the reproductive success. For example, aphids in summer produce enormous numbers of offspring by asexual reproduction and so, at this time, we can say they are maximising their fitness. We can compare this with their sexual reproductive phase before winter when genetic recombination provides more flexibility but fewer survivors, hence reduced fitness. We may not say 'fitter individuals will, on average, leave more offspring' because this is a circular statement equivalent to 'individuals that leave more offspring will, on average, leave more offspring'.

So, what is it about these fitter individuals that actually enables them to leave more offspring? Well, we can argue that such individuals must be better adapted. That is to say, some individuals have more favourable traits and are therefore better able to withstand the pressures of selection. Since they are able to withstand the selection pressures, they produce more offspring (that is, they are more fit). So the genes that these better-adapted individuals carry will be better represented in the next generation than the genes of the less well-adapted, and hence less fit, individuals.

SAQ 2 Explain the meaning of the term fitness.

1.2.2 The abiotic and biotic environments

The theme of this unit is survival, and so one question we can ask is 'What determines whether or not an organism can survive in a particular environment?' The most fundamental aspect of the environment is the set of physical and chemical conditions prevailing. Does the temperature get too high or too low? Is there enough light? Are there sufficient soil nutrients? Is the site too exposed to wind? These are the sorts of questions we start asking when looking at the physicochemical or **abiotic environment**. Of course, different organisms will have different tolerances, and we can start to define for individuals and for species lethal limits and preferred ranges. Remember these abiotic factors will not just determine whether an organism can survive in an environment; they may influence its fitness (reproductive success) and thus the genetic composition of future generations.

Clearly then, an organism must be adapted to the physicochemical or abiotic environment to survive. Just how it lives, the details of its ecology, may be influenced by the presence of other living organisms, the biotic factors. Relationships, both beneficial and damaging, will constitute the **biotic environment**. These may include various feeding relationships (including commensalism, mutualism, parasitism) as well as other sorts of interactions like pollination, fruit dispersal by animals, trampling and the secretion of chemicals by plants to prevent others growing nearby.

Finally, it must be said that the distinction between abiotic and biotic factors is not quite so clear-cut. It should be obvious that many organisms affect other organisms by altering the physical conditions, sometimes very locally, even on a micro scale. For instance, plants with root nodules, like clover, may raise the nitrogen level of the soil, whilst large animals can affect soil nutrients in their feeding and latrine areas. The changing nutrient level of the soil will influence the plant species composition it can support and, in turn, the animal communities too. (Remember from section 5 of unit 9, *Ecology*, that a community is all the biotic components, or all the living organisms, in an ecosystem.) In the extreme case of endoparasites, the biotic environment (that is the host) may constitute the whole of the abiotic environment too, with temperature, pH, nutrient status, and so on, all under the control of the host organism.

In other words, there is a two-way interaction between the biotic and abiotic factors, each influencing the other.

SAQ 3 Decide whether the following are abiotic or biotic factors:

(*a*) competition, (*b*) dry conditions, (*c*) light intensity, (*d*) predation, (*e*) pathogens, (*f*) evaporative cooling.

1.3 Group versus individual selection

We have seen that selection pressures operate on individual organisms. The traits, including behaviour, carried by that individual may or may not be well adapted to the conditions (biotic and abiotic) in which it finds itself. If it is well adapted, then the animal should prove fit (that is be reproductively successful) and thus transmit its genes to the next generation. Equally, non-adaptive behaviour should reduce the fitness of the animal and thus tend to be selected against. In view of this, since Darwin, biologists have had a problem in trying to explain altruistic behaviour.

Altruistic behaviour is behaviour which appears to benefit other members of the population but may actually confer a disadvantage to the individual behaving altruistically. For instance, a blackbird giving the warning 'pink-pink' call may benefit others in the vicinity but draw attention to itself, thereby risking predation. Also, in social insects, like the honey-bee, individuals devote themselves to communal work on the hive with no prospect of reproduction; indeed, in some termite species some castes may literally commit suicide in defence of the rest.

Since altruism appears to benefit groups, it was, at first, tempting to argue that those groups which manifested altruism were better adapted than those without and would prove fitter and so be selected for. However, no satisfactory mechanism can be postulated for group selection; natural selection always operates through the selection of individuals. So if group selection must be discarded, how else can we explain altruism?

Different solutions to the problem of altruism have been proposed, but **kin selection** is the theory that seems to explain most altruism. Inspection of altruistic behaviour shows that it is prevalent in groups of related individuals. Since close relatives carry similar genes, their altruistic behaviour would reduce the fitness of the individual but would promote the survival of other individuals which, in turn, would carry the trait, or tendency towards, altruistic behaviour. Hence, altruism would be promoted. J.B.S. Haldane once said that on this basis, he should lay down his life for two brothers or eight cousins!

Of course, there are problems. Firstly, it has proved difficult to demonstrate that genes for altruism exist. Secondly, some people have wanted to use a biological explanation of this type to account for human altruism. It seems fair to say an attempt to account for human ethics purely in terms of biology will be inadequate, if not downright misleading.

SAQ 4 Define altruism.

SAQ 5 In what way might altruism reduce fitness?

1.4 Short-term versus long-term pressures on survival

We saw in section 1.2.2 that abiotic (physicochemical) factors will impose limits on the distribution of organisms. Naturally, these abiotic factors will fluctuate, thereby causing problems for the immediate short-term survival of the individual but also conceivably for the long-term survival of the species. So what strategies are available to overcome such fluctuation, both within a generation (that is, short-term) and between generations.

There is an array of strategies to overcome short-term changes. We will examine a few. Many short-term changes are regular, with a daily, seasonal or lunar periodicity. In such cases, organisms may show changes in their physiology. For instance, many plants show changes in water levels and build up anti-freeze chemicals in their cells to withstand winter temperatures.

In mammals many events and processes are under the control of a daily or **circadian rhythm**. For instance, in humans sleep and urine production are rhythmic. We excrete more urine in the morning and at midday than at night. Processes under the control of a circadian rhythm may be linked or integrated. So readjusting one's sleep schedule will, for instance, cause a change in the pattern of excretion of potassium.

Sometimes, the change of environmental conditions may be so severe that it is beyond the normal tolerance range of an animal. Hibernation and the summer equivalent, aestivation, are possibilities open to some species, while certain insects encountering poor weather conditions may reduce their metabolic rate and enter a condition known as **diapause**. Other ways of temporarily 'shutting down' include encystment in *Amoeba*, seed dormancy in plants and egg dormancy in *Chirocephalus*.

In the event of moderate non-seasonal changes in an environmental parameter, the organism may undergo a limited internal physiological adjustment, known as **acclimation**. For example, certain tropical fish species, if kept for a time at lower temperatures, will withstand more cooling before death than those acclimatised at a higher temperature.

A more sophisticated way to cope with some fluctuation in the environment is **homeostasis**. This means keeping the internal state more or less constant. For example, mammals and birds can maintain their periods of activity independent of the prevailing temperature. Thus, in summer, the dawn chorus will start each day at approximately the same time, but on a cool morning much of the insect life may remain sluggish until later in the day. Remember, homeostasis may be achieved both by physiological and by behavioural activity; for example, many lizards keep their body temperature more or less constant during the day by appropriate behaviours like basking or seeking shade.

Not all organisms will necessarily employ all these strategies (homeostasis, circadian rhythms, dormancy, acclimation and so on) though some will. But no matter how good it is at withstanding short-term changes, an organism still has limits within which it can survive. These may be lethal limits such as a maximum temperature, a minimum light level or necessary oxygen concentration. Before lethal conditions exist, critical limits may be encountered. (For instance, the temperature may get so low that an animal may not die but is unable to move.) Even before the critical limits are reached, fitness may be impaired. How then are long-term gross changes in the environment dealt with? The answer is quite simply through two processes acting together. These processes are the creation and transmission of genetically inherited variations and the action of natural selection upon them. The competitive struggle for natural resources in the face of selection pressures ensures that only the adequately adapted organisms survive. The results of these changes constitute the process of evolution.

SAQ 6 What is homeostasis?

SAQ 7 Define natural selection in your own words.

SAQ 8 List the factors which contribute to the evolutionary process.

Survival in the short- and long-term will be considered as separate sections in this unit.

1.5 References and further reading

Neo-Darwinism by R.J. Berry, Studies in Biology No. 144.
Evolutionary Principles by Peter Calow.
The Selfish Gene by Richard Dawkins.

Section 2 Regulatory mechanisms (1) The control of the internal environment – homeostasis

2.1 Introduction and objectives

Living systems are not static, nor is the environment in which they live. This section discusses how organisms minimise the disruptive, and exploit beneficial, environmental factors which affect them, and also the mechanisms which enable them to do this.

After completing this section you should be able to do the following.

(*a*) Distinguish between the terms conformer and regulator.

(*b*) State the general principles of homeostasis.

(*c*) Distinguish between the types of equilibrium and feedback mechanisms.

2.2 Historical aspects

One of the most interesting features of physiological systems is that they are able to keep on working in a range of external conditions (abiotic factors). Claude Bernard, the great French physiologist, noted this over a hundred years ago and wrote 'it is the fixed state of the internal environment which is the condition of free and independent life ... All the vital mechanisms, however varied they may be, have only one object, that of preserving constant the conditions of life in the internal environment.'

What do we mean by the phrase internal environment or *milieu intérieur* as Bernard called it? We can take it to mean the system of fluid-filled spaces which immediately surrounds the cells of mammals. This extracellular fluid includes blood, lymph and the interstitial or tissue fluid. It is via the tissue fluid that food and oxygen are supplied to the cells and waste produces are eliminated. It is the blood which conducts gases and nutrients to the tissue fluid and removes waste products from it (see figure 1).

1 How tissue fluid is formed

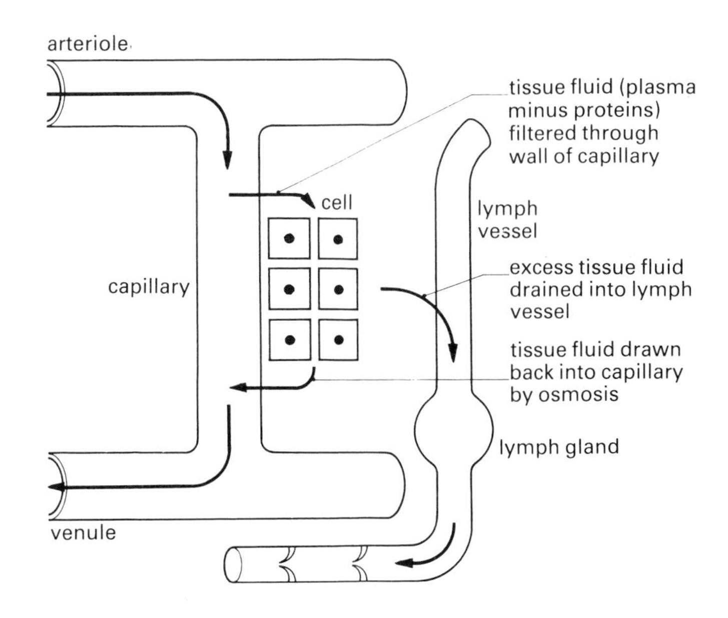

SAQ 9 What was the main observation of Bernard?

The physicochemical environment of a cell is determined by the composition of the extracellular fluid. Such factors as water, macromolecules, electrolytes, gases, pH, temperature and pressure will contribute to its composition.

Clearly, the metabolic activities of the cells will bring about changes in the physicochemical environment. Substances will be added to and withdrawn from the fluid and the amount of each constituent will therefore vary. Survival of the cells (and thus the whole organism) is made possible by mechanisms which regulate and keep more or less constant the composition of the extracellular fluid.

SAQ 10 What is the main factor which causes disturbances in the composition of the extracellular fluid?

The process of regulating the internal environment is an example of **homeostasis**. This term was first coined in 1929 by the American physiologist Walter B. Cannon, and is derived from two Greek roots, 'homeo' meaning similar, and 'stasis' meaning standing.

2.3 Conformers and regulators

Remember the internal environment will be influenced not only by the activities of the cells it surrounds but also by changes in the external environment (external abiotic factors) (figure 2). Some biologists think that the ability of animals to regulate their internal environment (that is to maintain homeostasis) is the essential factor which enabled them to live in physiologically hostile environments, though this is perhaps a slight oversimplification.

2 Influences on and by the internal environment

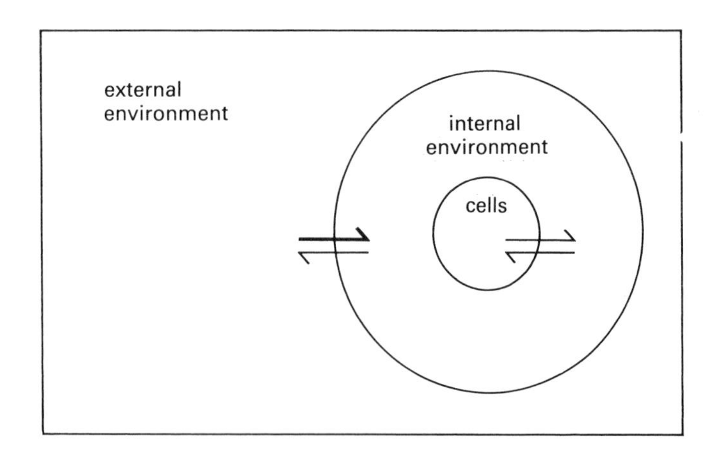

Different groups of animals deal with changes in their external environment in different ways but we can identify two broad categories: conformers and regulators. **Conformers** are organisms whose internal conditions fluctuate directly with external abiotic factors, for instance the osmotic concentration of marine invertebrates or the body temperatures of most insects. For conformers, the limits of tolerance will be imposed directly by the abiotic factors of the environment; for example marine invertebrates that stray into brackish water may die, and insects in low temperatures will be unable to move (a factor which nature photographers frequently take advantage of!) It is too easy to say that **regulators** (animals with homeostatic mechanisms) will escape tolerance limits because the abiotic factors do not influence the internal environment. Regulators also have tolerance limits. External conditions may reach limits that the homeostatic mechanisms cannot cope with; for example hypothermia in elderly humans who lack adequate heating in cold weather.

Remember that, as we saw in the last section, for both regulators and conformers the tolerance limits may be influenced through acclimation or evolution. Nevertheless, in general, we can say that regulators will have wider tolerances than conformers.

SAQ 11 Figure 3 depicts the variation of blood concentration of two crabs in the face of changing external conditions. With respect to blood concentration, (*a*) which of the two crabs is more of a conformer and which is more of a regulator, and (*b*) which of the two will be able to tolerate a wider range of external concentrations, the conformer or the regulator?

SAQ 12 What advantage has an animal which can maintain homeostasis over one that cannot?

We shall now go on to consider some of the features of homeostatic mechanisms.

3 Concentration of blood of two crabs as a function of the external medium. *Telphusa* is a freshwater crab, *Maia* is a sea crab

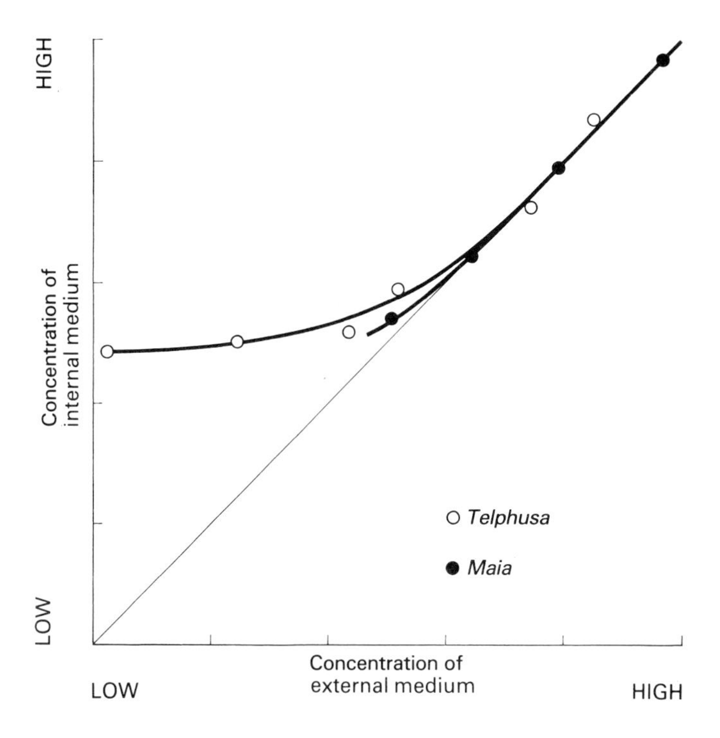

2.4 Equilibrium

Homeostatic mechanisms are self-contained such that they do not require any external information in order to maintain the extracellular fluid in a state of equilibrium. This means that the quantities of the components of the fluid remain within narrow limits. Because the actual molecules are being exchanged, removed and replaced, we call this **dynamic equilibrium**. This state of dynamic equilibrium is rather like the level in a reservoir: water is constantly being replaced, but the level remains the same (see figure 4).

4 The maintenance of water level in a reservoir by a dam

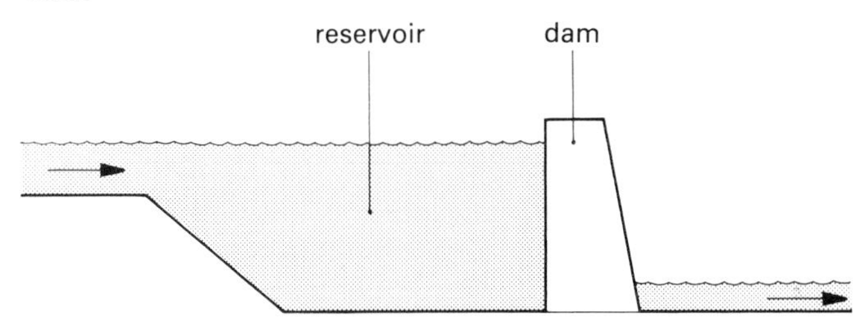

SAQ 13 What is meant by the term dynamic equilibrium? What do you think the term static equilibrium would mean?

2.5 General principles of homeostasis

In order for adjustments to be made to changing conditions, there must be three components to a homeostatic mechanism:

1 Input – the 'desired' or optimal physiological state must be known by the comparator.
2 Comparison – the 'desired' state must be compared with the existing state.
3 Output – compensatory action which is proportional to the difference between the existing and desired levels must then be taken by the effector muscles and glands.

An analogy may help you to understand this. A typical homeostatic mechanism acts rather like a person riding a one-wheeled cycle. The rider always has the tendency to fall off, so he must continually adjust the position of the cycle underneath. To do this effectively, he must, at all times, be aware of the position of his body relative to the position and movement of the cycle. His central nervous system is fed this information by the eyes, touch receptors, and vestibular mechanisms (the organ of balance in the ear). Compensatory movements of the skeletal muscles are commanded by the cerebral cortex, after the information from the sense organs has been processed by a separate area.

SAQ 14 What are the three main components of a homeostatic mechanism?

In much the same way, information regarding any changes in the state of their extracellular fluid is fed back to the relevant organs of homeostasis, the effector muscles and glands, which make the necessary adjustments.

The way this operates in mammals is partially summarised in two diagrams shown in figures 5 and 6.

5 Homeostatic control of the osmotic pressure of extracellular fluid

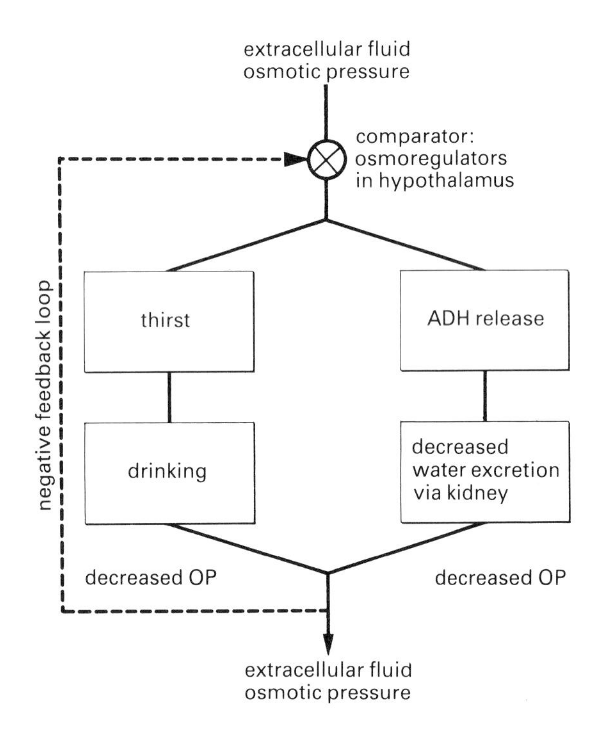

6 Homeostatic control of the volume of extracellular fluid for SAQs 15–17

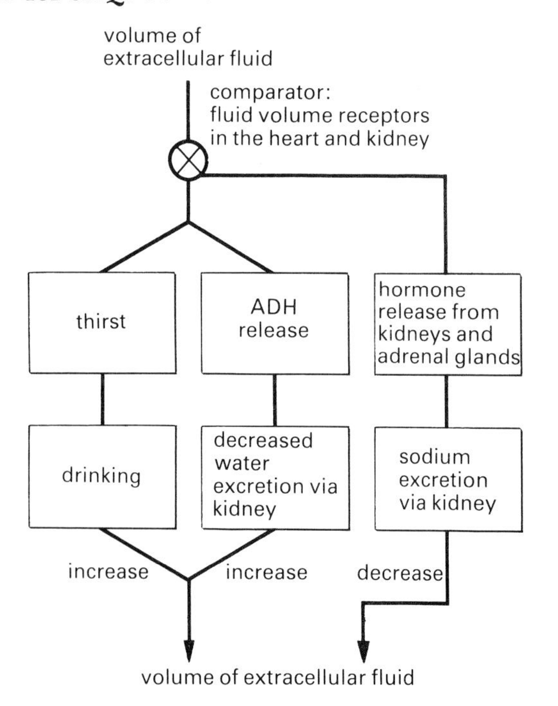

The osmotic pressure and volume of the extracellular fluid are monitored continually by the brain, which compares the monitored levels with a 'desired' or optimal physiological level. Differences between the monitored and desired levels are due to disturbances in the extracellular fluid composition and volume. In the main, these disturbances are due to the metabolic activities of cells. However, more serious disruptions may result from such disorders as vomiting and diarrhoea, or excessive water intake. The effector mechanisms which offset changes in pressure are shown in figure 5.

The new value for osmotic pressure produced by these effectors is continuously compared with the desired level. This comparison, or feedback of information is essential for the functioning of any homeostatic mechanism.

2.6 Feedback mechanisms

You will remember from section 2.5 that there are three parts to a homeostatic control system, namely input, comparison and output. In order to be self-regulating, an indication of the state of the output of the system must be 'fed back' into it. This feedback is used to compare the output with the desired level (figure 7).

7 A feedback loop

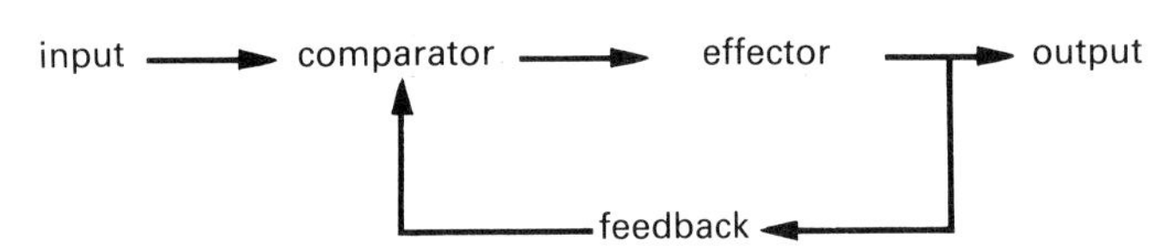

There are two types of feedback, negative and positive. **Negative feedback** occurs when a change from the desired level in one direction prompts a command for a change in the opposite direction. An example of negative feedback has been seen in section 2.5. Any change in the level of osmotic pressure leads to two separate correcting mechanisms being put into operation. This is shown in figure 5. When osmotic pressure gets above a certain level, the release of antidiuretic hormone (ADH) from the posterior lobe of the pituitary gland is stimulated. This hormone reduces the loss of water from the body via the kidney. At the same time, the same hypothalamic osmoreceptors stimulate thirst, which leads to increased water intake. As these two mechanisms, working together, bring about a lowering of osmotic pressure, the osmoreceptors are made aware of the new value. As a result of this, thirst/drinking stimulation and ADH release are adjusted accordingly. Osmotic pressure is therefore kept in equilibrium by negative feedback.

Similarly, the volume of extracellular fluid is compared by fluid volume receptors in the heart and kidney. Disturbances in volume are compensated for by the thirst and ADH mechanisms, and also by hormones released by the kidney and adrenal cortex. These hormones adjust the amount of sodium excretion. Information regarding the new extracellular fluid volume is then fed back to the receptors.

SAQ 15 Effector mechanisms exist to increase or decrease the volume of extracellular fluid. How many feedback loops will this system require?

SAQ 16 On a copy of figure 6 complete the diagram by drawing in the loop(s) giving negative feedback.

SAQ 17 Continuous regulation of blood sugar is achieved by negative feedback. Describe this, using a flow diagram similar to those in figures 5 and 6.

Positive feedback is said to operate when a disturbing influence operating in one direction is converted into a command to enhance that trend. An example of this can be seen in the hormonal control of the reproductive system in mammals.

You will remember from unit 4, *The continuity of life*, that follicle stimulating hormone (FSH) causes the maturation of the Graafian follicle. During this time luteinising hormone (LH) stimulates the follicle to produce oestradiol (an oestrogen). Oestradiol has a negative feedback effect on FSH, keeping its level low and, gradually, as the follicle matures and the oestradiol level builds up, even depressing FSH levels. Eventually the level of oestrogen (oestradiol) reaches a critical level where it ceases to have a negative feedback influence; rather it stimulates FSH and LH production up to the point of ovulation. This stimulation phase is an example of positive feedback and over the course of two or three days the FSH and LH rise to many times their previous concentrations in the plasma.

Following ovulation a number of changes now occur. The follicle, now ruptured, develops into the corpus luteum which secretes progesterone and low levels of oestrogen. Progesterone has a direct negative feedback effect on FSH and LH. The progesterone also seems to have the effect of inhibiting the positive feedback effect of high oestrogen levels. These two effects rapidly reduce the circulating levels of FSH and LH as both oestrogen and progesterone quickly inhibit their release from the pituitary.

SAQ 18 In your own words, describe how the oestrogen/LH system provides an example of positive feedback.

Because of the double effect of progesterone (inhibition of FSH and LH production and prevention of positive feedback effect by high levels of oestrogen) it has been used in certain contraceptive pills. For instance, in combined oral contraceptives both progesterone and oestrogen can be used to inhibit follicle development and hence ovulation. Such pills require a seven-day break to induce a sort of menstrual bleed but other combinations of the two hormones have been and are used.

Positive feedback is generally used in biological systems to produce an explosive or regenerative event. Often this generates the rising phase of a cyclic event.

SAQ 19 Explain how the upstroke of the nerve action potential provides an example of positive feedback.

2.7 References and further reading

Homoeostasis by R.N. Hardy, Studies in Biology No. 63
Biology: a functional approach by M.B.V. Roberts.
Biological Science Book 2 by N.P.O. Green, G.W. Stout & D.J. Taylor.
Reproduction in Mammals by C.R. Austin and R.K. Short (eds.).
Physiology of Mammals and Other Vertebrates by P.T. Marshall & G.M. Hughes.

Section 3 Regulatory mechanisms (2) The response to environmental changes

3.1 Introduction and objectives

This section is about the physiological and behavioural strategies involved in survival, and how these are modified in order to cope with changing conditions.

After completing this section you should be able to do the following.

(*a*) Describe the homeostatic control mechanism for homeothermy.

(*b*) Outline the physiological process of hibernation.

(*c*) Define heterothermy and give examples.

(*d*) Explain how aquatic homeotherms are able to maintain their body temperatures.

(*e*) Describe the response to short- and long-term exposure to conditions of low oxygen, for example at high altitude.

(*f*) Explain the respiratory and circulatory adaptations of aquatic air breathers for diving.

(*g*) Describe the survival value of fever.

(*h*) Outline the behavioural responses of homeotherms and poikilotherms to extremes of temperature.

(*i*) Understand the importance for survival of migration and social organisation.

3.2 Physiological strategies

3.2.1 Homeothermy and heterothermy

You will remember from unit 6, *Response to the environment*, that homeothermic animals (birds and mammals) have physiological, as distinct from behavioural, methods for maintaining their body temperatures at a stable level which is independent of the temperature of their external environment. In a competitive environment, being a homeotherm has considerable survival value. By maintaining a steady body temperature at a level at which the metabolism of cells works best, homeotherms are free to exploit a wider range of habitats all the year round.

8 Homeostatic maintenance of homeothermy

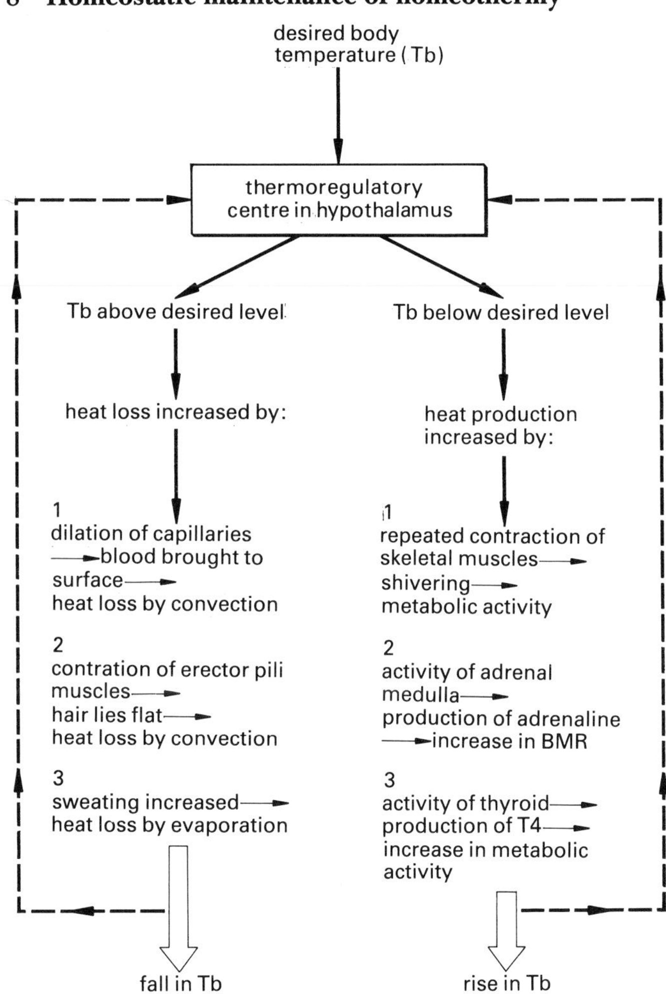

Homeothermy is maintained by a homeostatic feedback mechanism (see figure 8). In order to be able to exploit the habitats in which the more extreme conditions prevail, some homeotherms have to compromise their normal homeostatic mechanisms in order to survive. There are several strategies for this.

3.2.2 Hibernation (See unit 6, *Response to the environment*, section 6.8)

Most animals that hibernate are small and have a high surface area to volume ratio. They give up the struggle to maintain a high constant body temperature as the external temperature falls, and so are able to cut down on the 'cost' of keeping warm. They use less fuel and so conserve their energy reserves. The mechanism which enables them to survive in this state is not a failure to maintain body temperature, but a well-regulated physiological process. The yearly hibernation cycle seen in such animals as bats, dormice and hedgehogs is influenced by the photoperiod and by the endocrine cycles within the animals.

The animal's body temperature decreases as the ambient temperature decreases. At a point approaching 0°C, one of two things happens to the hibernator. Either it arouses, and returns to the active state, or heat production is switched on and maintains the body temperature at a steady low level even if the ambient temperature falls below freezing-point. In this way damage of tissues due to freezing is prevented, and also fuel is saved.

This can be seen in the West Indian hummingbird, *Eulampis jugularis* (figure 9). When the bird is torpid, its body temperature approaches air temperature. If this falls below 18°C, the body temperature does not fall any further, but is maintained at a constant 18–20°C.

SAQ 20 In *Eulampis*, heat production must be increased as the ambient temperature drops further and further below 18°C. What shape would a graph be of heat production against environmental temperature for *Eulampis*?

9 Oxygen consumption in *Eulampis jugularis*

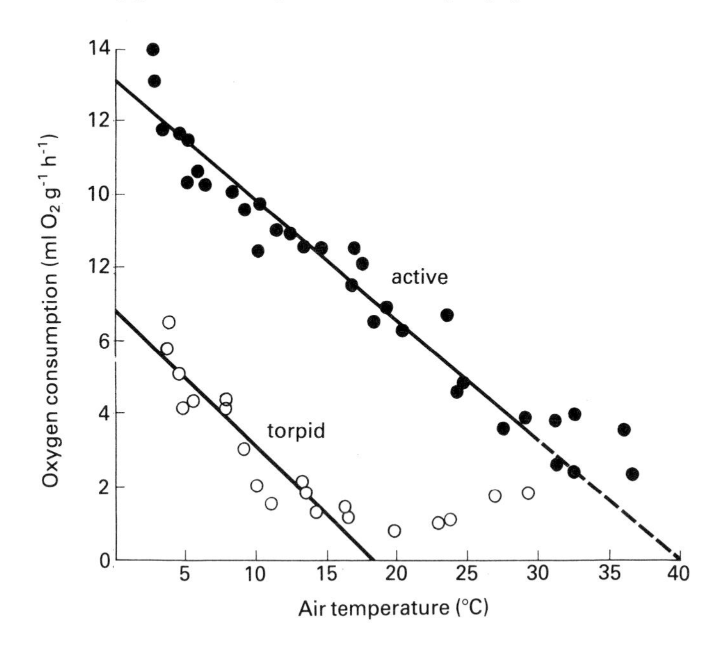

3.2.3 Heterothermy

Animals that, in normal conditions, have a well-regulated constant body temperature but in extreme conditions can vary their body temperature, are known as **heterotherms.** The bactrian camel normally inhabits hot and arid environments, and can allow its body temperature to change over a considerable range in order to survive these conditions. The increased daily temperature of camels in the Sahara has been shown to vary between 34 and 40°C. The greatest problem for survival facing the camel in these conditions is that of conservation of water. Varying its temperature in this way helps the camel to cut down water loss due to evaporation. By letting its temperature rise it avoids having to use the evaporation of water to lose heat to maintain a lower constant temperature. As the temperature of the air drops at night, the heat which the camel has generated or absorbed during the day is lost to the cooler night air by radiation. Just as hibernation is a survival ploy of small animals, so heterothermy, as shown by the camel, can only be used to survive extremes of temperature by larger animals. This is due to their relative surface area to volume ratios.

SAQ 21 What advantage might each of the following features of a camel confer on it when in the desert?

(*a*) Does not pant
(*b*) Moderate sweating
(*c*) Concentrated urine
(*d*) Variable body temperature
(*e*) Heavy coat
(*f*) Ingests a great volume of water at one draught

3.2.4 Regional heterothermy

Another physiological stratagem employed by some marine heterotherms involves allowing some parts of the body to become significantly cooler than the thermal centre of the body. The heat conductivity of water is about 25 times as great as that of air, so what might produce a large rise in air temperature would only give a relatively small rise in water temperature. Most marine mammals have a thick insulating layer of fur or blubber, or both, and also have a low surface area to volume ratio. However, in order to move in water they must have flukes or flippers, and these are generally poorly insulated. These appendages have a good blood supply, so body temperature would be reduced without a good heat conservation mechanism.

The flippers and flukes of dolphins and whales have a countercurrent heat exchange system whereby the core temperature can be maintained whilst the temperature of the appendage is allowed to drop (figure 10). (The topic of countercurrent heat exchange systems was covered in section 6.6 of unit 6, *Response to the environment*.)

10 Countercurrent heat exchange system in the flipper of a dolphin

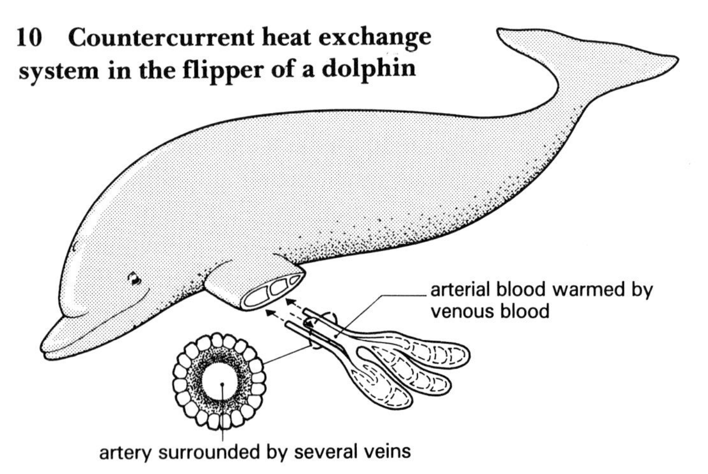

Each of the arteries is surrounded by veins. The returning venous blood cools the outgoing arterial blood which gives up its heat to the returning blood. The venous blood is therefore warmed before re-entering the body, and the arterial blood cooled before entering the flipper (figure 11).

SAQ 22 Explain what is taking place in this model of a countercurrent heat exchanger.

11 Model of a countercurrent heat exchanger

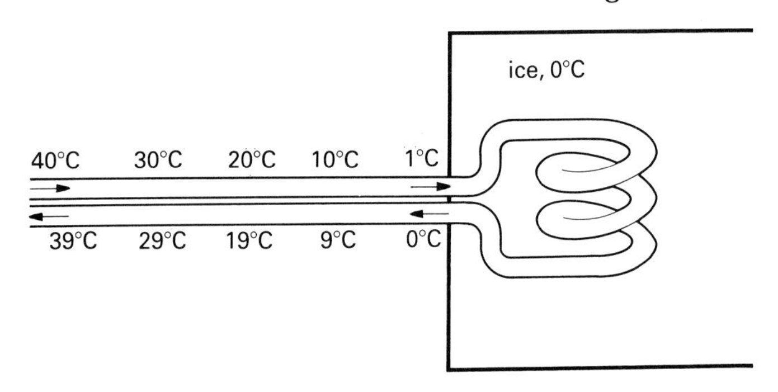

3.2.5 Adaptations to decreased oxygen availability due to high altitude

At high altitudes the atmospheric pressure is lowered and this means that the partial pressure of oxygen, pO_2, is lower than at sea level. In humans, a reduction of oxygen in the air, or **hypoxia,** results in a decrease in blood pO_2. Our response to this is to increase the rate of breathing. This is known as **hyperventilation**. The effect of this is that carbon dioxide elimination is increased and the level of blood carbon dioxide is decreased. In turn, this brings about an increase in the pH of body fluids, a condition known as **respiratory alkalosis**.

If the conditions of reduced oxygen persist, after about a week the rate of ventilation of the lungs and the pH value of the body fluids are returned to normal levels. This is due to the elimination of bicarbonate by the kidney.

Over a much longer term humans, and indeed most vertebrates, respond to low oxygen conditions by increasing the number of red blood cells, and the total blood haemoglobin content. This increases the oxygen-carrying capacity of the blood.

SAQ 23 What are the responses of humans to hypoxic conditions?

SAQ 24 How does respiratory alkalosis occur?

Animals, such as the llama, which normally live at high altitudes have haemoglobin with a much higher affinity for oxygen. This means that the haemoglobin can become fully saturated with oxygen at relatively low atmospheric oxygen levels. This is not the case in humans, in whom the ability of the haemoglobin to take up oxygen at lower partial pressures remains substantially unchanged.

SAQ 25 Plot the curve for the llama on a copy of figure 12.

12 Oxygen dissociation curve for a human

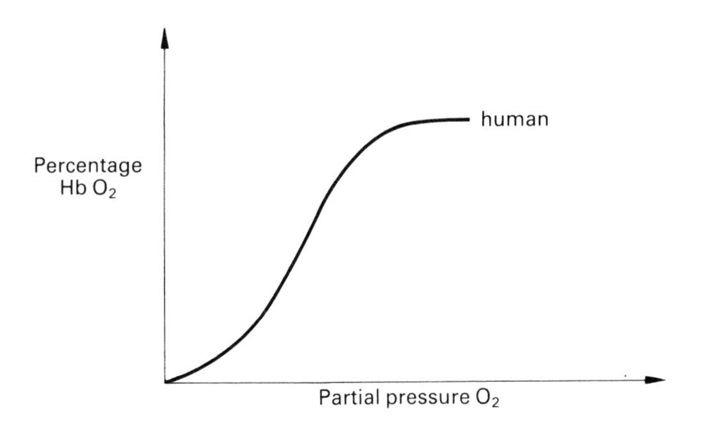

In llamas, the normal tissue functions occur at low pO_2 values. This is true of zoo-bred llamas kept at sea level, so it is a true adaptation to hypoxia.

3.2.6 Adaptation to decreased oxygen availability in diving mammals

Anoxia is the complete deprivation of oxygen. In air breathers, this occurs during asphyxia, but the diving mammals and birds frequently endure periods of anoxia in the normal course of their lives. The common response of all mammals, both aquatic and terrestrial, to anoxia is a decrease in heart-beat rate **(bradycardia)** and some redistribution of blood

13 Heart rate in a seal during a dive

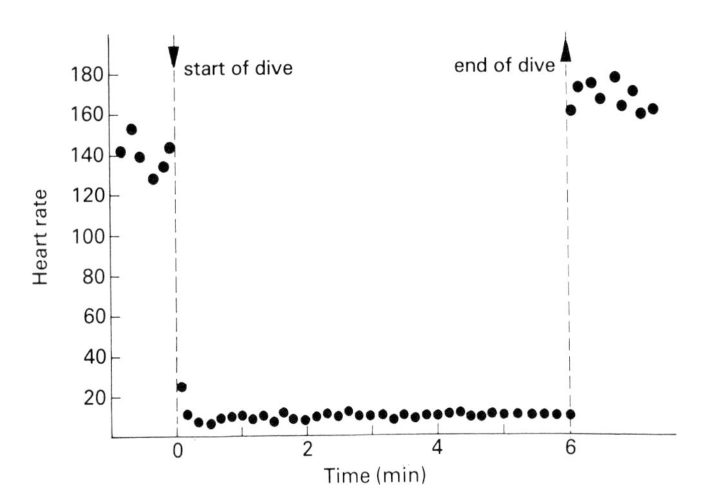

within the body. This response is exaggerated in the diving mammals (such as seals and dolphins) and birds (for example, penguins and ducks) (see figure 13).

The maximum diving time for humans only exceeds 2 min in very rare circumstances, but seals can remain submerged for 15 min or more. How do they manage to make their oxygen supply last for so much longer than a human?

The most common adaptation seen in successful diving species is an increase in the blood volume. This, together with a slight increase in the oxygen-carrying capacity of the haemoglobin, ensures that an increased amount of oxygen can be carried in the blood.

The meat of diving mammals is characteristically a much deeper red than that of terrestrial species. This is because the muscle tissue has a greater myoglobin content. The significance of this for diving is that the myoglobin facilities oxygen transport from the blood into the mitochondria in actively respiring tissues. The muscles can thus continue to contract aerobically for a longer period than in terrestrial forms during anoxia. However, these muscles can continue to function anaerobically with the production of lactic acid when oxygen supplies run out.

In consequence, the oxygenated blood can be redistributed to those organs for which oxygen is essential, that is the heart and central nervous system. Blood supplies to the viscera and voluntary muscles is reduced, thus ensuring that lactic acid produced by anaerobiosis in those tissues is prevented from entering the general circulation during the dive.

As a result of these adaptations, the oxygen is used up much more slowly in air-breathing animals adapted for lengthy dives. The actual length of a dive has been shown to correspond with the time taken to fully deplete the oxygen in the blood.

The natural action of a human about to make a dive would be to take a deep breath, but diving mammals exhale before diving. When they dive deeply, the pressure of the water on their thorax causes lung compression. Because the animal has expelled the air from its lungs, as the pressure increases, the alveoli collapse and residual air is forced into the bronchi and trachea, where gaseous exchange with the blood cannot take place. If the lungs were full of air, the increased pressure would force gases into solution in the blood via the alveoli. Decompression as the animal surfaced would result in the formation of bubbles of nitrogen in the blood, the equivalent of the 'bends' in humans.

Information regarding the levels of gases in the blood normally comes to the brain from the carotid and aortic bodies, resulting in an increase of ventilation during hypoxia. This response of the lung ventilation is suppressed in diving mammals by signals from the water receptors on the nose and face. These are stimulated in the presence of water, and inhibit inspiration during a dive.

SAQ 26 Summarise the respiratory and circulatory responses of the diving mammal.

3.2.7 Failure of regulation of body temperature – fever

It has been known since the time of Hippocrates (400 BC) that fever occurs in humans. The German physician Liebermeister, in the late nineteenth century, pointed out that body temperature in a patient with fever was regulated at a new higher level. He found that if his subjects were warmed or cooled, their thermoregulatory mechanisms returned their body temperatures to the new higher level. This is interpreted as an increase in the desired level of body temperature, or set-point.

Behavioural mechanisms too are adjusted in people with fever.

In figure 14, subjects who were in a warm bath at 40°C were asked to put their hands into gloves outside the bath. The gloves were warmed or cooled to known temperatures. Control subjects (those with normal body temperature) found the most pleasant glove to be warmer than the water bath. As the body temperature of the subjects rose, increasingly cooler gloves were chosen. Subjects with fever showed precisely the same responses, except that the cooler

14 Change in preferred glove temperature in fever

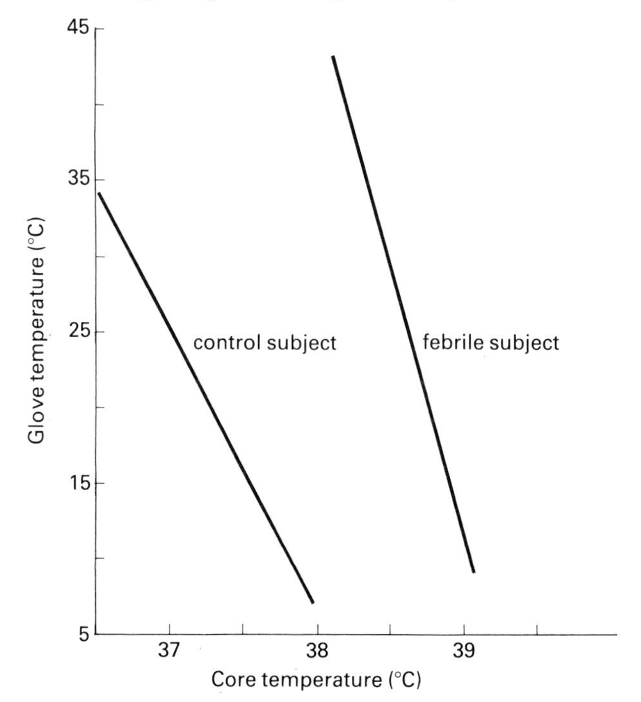

gloves were first chosen at a body temperature a full 1°C higher than for the subjects without fever.

The hypothalamus seems to be responsible for choosing this set-point of body temperature. This part of the brain contains neurons which are temperature sensitive, and these form the central part of the thermoregulatory system. Neurons which are 'warm sensitive' increase their firing rate when the temperature of the hypothalamus is increased. Experiments have shown that these neurons show a decrease in activity for a comparable hypothalamic temperature in rabbits which have a fever. The opposite effect is seen in the 'cold sensitive' neurons. These observations have led to the suggestion that these neurons form the physical basis of the change in set-point seen in mammals with fever.

The raised set-point seen during fever has been shown to have survival value. Animals which develop fever during disease survive better than those which do not. It has frequently been reported that drugs which reduce fever (antipyretics) do not, in fact, increase survival rate and may actually increase mortality.

The survival value of fever during disease is not restricted to mammals, and has been demonstrated to occur even in reptiles, which regulate their body temperature solely by behavioural means. The fact that fever seems to exist in the lower vertebrate classes suggests that it has an ancient evolutionary origin. This argues strongly for fever being an important adaptation for recovery from disease.

SAQ 27 Figure 15 shows the body temperature of a lizard which was regulating its own body temperature in a varied environment. The lizard was in a wooden box, one end of which was 30°C and the other was 80°C. The peaks on the graph (high set-points) show when the lizard moved to the cooler end, and the troughs (low set-points) show when it moved to the hotter end. Enter a line on a copy of the graph to show what you think would be the body temperature of the lizard if it had fever. Explain your graph.

15 Body temperature of a lizard

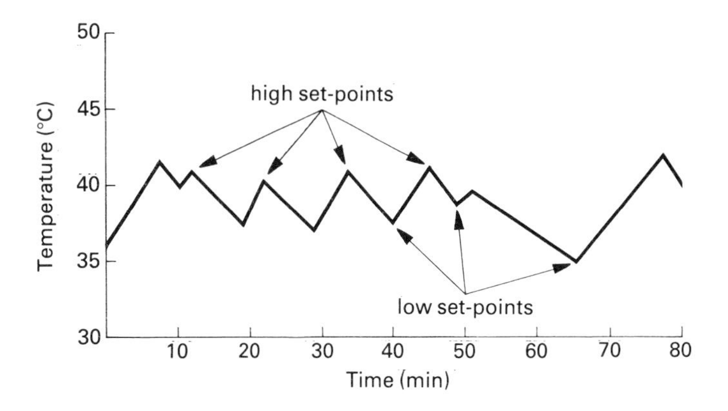

3.3 Behavioural strategies

The following paragraphs discuss how the behaviour of animals helps them to survive in adverse conditions. It must be stressed, however, that underlying every apparently behavioural stratagem is a physiological mechanism. It is not possible to separate these functions.

3.3.1 Homeotherm behaviour

If, because of its size or way of life, a homeotherm is unable to take shelter from extremes of environmental temperature, it can substantially change its heat gain or loss by behavioural means. The surface of the body is the major site of heat exchange so, if the animal can alter its surface area to volume ratio, it will be able, to some extent, to regulate its body temperature. Camels in the desert huddle together in groups during the day. This way they are able to reduce the surface area which is exposed to sunlight, and increase the surface area which is exposed to another camel which itself is cooler than the surroundings. Likewise, juvenile penguins in the Antarctic huddle together in creches, but this time it is to conserve heat.

SAQ 28 Give some more examples of this type of huddling behaviour.

If a homeotherm can find a different thermal environment, then it can avoid extremes of temperature. Many small desert animals are nocturnal (when air temperatures are low) and live in burrows during the day. Burrow temperature rarely exceeds 30°C even though the desert surface temperature may reach 80°C.

SAQ 29 Look at figure 16. What is the percentage difference in annual range between air temperature in a burrow 1.5 m deep and air temperature at the desert surface?

15 Diagram showing desert soil temperature variation at different depths

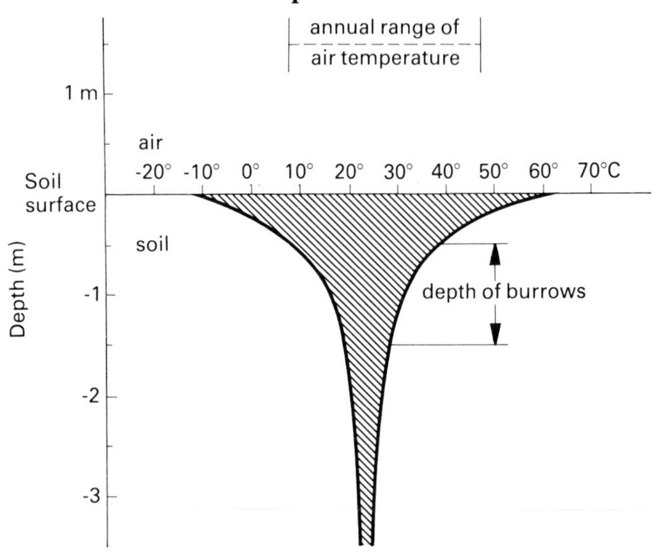

3.3.2 Poikilotherm behaviour

Poikilotherms rely much more on behavioural means of regulating their body temperatures than on purely physiological mechanisms. Reptiles, such as lizards, maintain a fairly narrow range of body temperatures (varying it by 4–6°C) by basking when their body temperature is too low, and sheltering when it is too high. Lizards can, by basking, maintain their body temperature 30°C above air temperature, which may fall below freezing. However, this is an example of a behavioural response being backed up by a physiological mechanism. Heat is rapidly and effectively redistributed from the exposed surface to the rest of the body.

The heat in some environments is so great that reptiles can gain too much heat. Under these circumstances, the animals wriggle their bodies into the cooler regions of the sand.

SAQ 30 What advantage does each of the following characteristics confer on a reptile in the desert?
(*a*) Basking
(*b*) Limited body contact with the ground (such as lizards lifting their feet, and sidewinder snakes)
(*c*) Scaly skin
(*d*) Concentrated urine

3.3.3 Social organisation and survival

The social organisation of the honey-bee enables the temperature of the hive to be kept constant in the face of changing outside temperatures. Honey-bees are able to generate heat by repeated contractions of their flight muscles, without actually flying. If environmental temperatures are too low, worker bees huddle in clusters and maintain the temperature of the cluster at 18–32°C, even at outside temperatures as low as −17°C. The surface area to volume ratio of the clusters helps to conserve the heat produced.

If hive temperature rises too high, the bees spread water over the comb and then fan their wings to cause evaporative cooling of the hive.

In both ways, hive temperature is kept suitable in a much wider range of environmental temperatures.

3.3.4 Migration: survival by redistribution

One of the most common behavioural strategies for survival is migration. This way animals can avoid extreme variations in environmental conditions by not being there when the adverse conditions occur. Migration, however, is a more positive behaviour than avoidance, for two reasons:

1 animals that migrate are able to anticipate unfavourable conditions;

2 migrants can relocate to environments that are positively favourable to survival.

Migratory behaviour, in fact, is so refined as to be concerned with the exploitation of not one, but many niches appropriate to the needs of the migrants at various times of their lives. This is more sophisticated than just being able to predict the onset of adverse conditions.

That the seasons are changing in the temperate zone and perhaps becoming less hospitable is conveyed to migrants by changes of day-length, temperature and rainfall. However, this type of information may not be readily helpful to the transequatorial migrant. Clues received in one hemisphere may not give accurate information about conditions in the other. European warblers, which breed in Europe, spend their winters south of the equator. They have endogenous annual cycles which predict the important events of their life-cycles (such as migrating, moulting, breeding) at the appropriate times for their survival in the Northern Hemisphere, even though they winter in areas in the Southern Hemisphere lacking predictive information.

SAQ 31 The migratory behaviour of birds has been the topic of much debate and research. Put forward your own ideas for the phenomenon of migration under the following headings:
(*a*) ecological explanation,
(*b*) historical explanation,
(*c*) internal physiological explanation,
(*d*) external environmental explanation.

3.4 References and further reading

Animal Physiology – Adaptation and Environment by Knut Schmidt-Nielsen.
Desert Animals – Physiological Problems of Heat and Water by Knut Schmidt-Nielsen.
Bumble-bee Economics by B. Heinrich.

Section 4 Adaptation – evidence for evolution

4.1 Introduction and objectives

This section is about adaptation and its significance for evolutionary theory. The concept of adaptation itself is discussed. This is followed by a series of topics which link adaptation to evolution.

After completing this section you should be able to do the following.

(*a*) Give several different explanations of the term adaptation.

(*b*) List some of the ways adaptations are related to various aspects of the environment.

(*c*) Describe different types of adaptation.

(*d*) Explain what adaptive suites are.

(*e*) Define niche and explain how adaptation is linked to niche exploitation.

(*f*) List the main geochronological aeons and link these with the emergence of life.

(*g*) List the main geological eras and link these with the origin of some of the major plant and animal phyla.

(*h*) Explain the significance and limitations of interpreting the fossil record.

(*i*) Define the term extinction and give examples of different types.

(*j*) Define the term adaptive radiation and, with examples, explain its significance in evolution.

(*k*) Explain why evolution is no longer seen as a steady process.

4.2 Adaptation

You will recall from section 1 of this unit that the use of the term fitness has changed over the years and come to mean reproductive success. Accordingly, such statements as 'fitter individuals will tend to leave more offspring' are trivial and circular. So what is it about certain organisms that makes them reproductively successful? One possible answer is to say that they are adapted. So what does adapted mean? One definition of adaptation is 'conformity between the organism and its environment'. In other words, the organism, if it is adapted, is able to cope with all the various factors of its biotic and abiotic environments by using the various means at its disposal (genetic, physiological, behavioural and so on). Thus, being adapted might be taken to mean having that minimum set of characteristics to enable perpetuation of the species. But this is a rather weak sense of the word. Biologists often seem to want to say more than this when they use the words adaptive, adapted and adaptation. Biologists may say that the clustering behaviour of penguins or the hump of the camel or the stripes on a tiger are adaptations to their environment; in other words we can see clear advantages to the organism in having such characteristics. So, are all characteristics held by an organism more or less adaptive, or can some be neutral? Ernst Mayr, a seminal writer in this field, wrote 'If a given subspecies of ladybird beetles has more spots on the elytra than another subspecies, it does not necessarily mean that the extra spots are essential for survival in the range of that subspecies. It merely means that the genotype that has evolved in this area as a result of selection develops additional spots on the elytra.' This is one point of view. Nevertheless, a debate still exists regarding what proportion of an organism's characteristics can be considered neutral rather than adaptive or unadaptive.

Some workers confine the application of the term adaptive to the sum total of an organism's characteristics. Such a view clearly acknowledges that there are a whole variety of 'design constraints', many of which will conflict and so need to be traded off against one another. For instance, smaller herbivores need to balance their time spent actively feeding against the time they devote checking for predators. If overly wary, their feeding efficiency will suffer and, if unduly incautious, their particular genes may be even more rapidly withdrawn from the gene pool! (One can imagine all sorts of equivalent examples like why is the hump of the camel not even larger or the human's loop of Henle even longer.) A disadvantage of confining the term adaptive to sets of characteristics in this way is that one is in danger of finding that the word adaptive may lose its explanatory value. Let us see why. Suppose we ask the question 'Why is a species as it is?' We can no longer answer 'because it is fit' but now we want to say 'because it is adapted'. If, by this, we mean 'because it is adapted, that is it has features *x*, *y* and *z*' then we have to be clear what we mean by a characteristic being adaptive. On the other hand, if we mean 'because it is adapted, that is the sum total of its characteristics is adaptive,' then we must ask 'How do we know the sum total is adaptive?' The obvious answer to this is likely to lead to a hideous circularity as the only real evidence we have for the adaptiveness is the past fitness (reproductive success) of a species.

The crux of the matter seems to be that the concept of adaptation embodies a notion of improvement. Marjorie Grene has written that 'An organism is regarded as adapted if one can imagine its condition as an improvement over another possible condition that would be slightly less favourable.' Now, why is this important? It is important for several reasons, reasons that you may have to bear in mind as you go on to evaluate the evidence for evolution. Firstly, we must realise the difficulty of 'proving' a characteristic to be adaptive; the judgement that something is adaptive is more like a thought experiment 'What if it were other than it is?' Secondly, there is a problem about how we are to judge if something is an improvement and what concept of improvement we are to employ. The term improvement could have a variety of meanings such as (*a*) increasing the likelihood of producing offspring, (*b*) in the sense of specialisation, fitting in better to a niche, or (*c*) it could mean the emergence of new complexity, maybe a gene, an organ or a piece of behaviour.

In conclusion, then, we can see that the notion of adaptation is not entirely straightforward, though that is not to deny it is a useful concept, and we shall now go on to consider more carefully some types of adaptation.

Adaptations are related to various aspects of an organism's environment (see section 1.2). Let us consider four aspects of environment:

1 abiotic environment,
2 biotic environment,
3 deme: conspecific, local breeding or social group environment,
4 internal environment.

SAQ 32 Put the following adaptations into figure 17. Blubber in whales, white flashes around the tail of some antelopes, water receptors on the face of some seals, carnassial teeth in carnivores, elongated loop of Henle in the kangaroo rat, the haemoglobin of llamas.

17 Table for SAQ 32

Environment	*Adaptation*
abiotic	
biotic	
deme	
internal	

It can be seen from this SAQ that it is quite difficult to separate the various components of an animal's environment when it comes to deciding on any given adaptive feature. Of course, in order to promote survival and fitness, the organisms have to use adaptive features in a range of physiological, structural and behavioural ways. As we have seen in earlier sections of this unit, some compromises may be necessary in order to promote overall fitness for survival. In these circumstances, the observer may think that the characteristics are not working in an optimal way in terms of their adaptive functions.

J. Maynard Smith, a great writer on this subject, gives three types of adaptation.

Genetic adaptation. Genetic adaptations are those characteristics suiting an organism for life under certain conditions. These characteristics are also developed in all environments in which the organism itself develops. For example, coat colouration in tigers is best suited to its natural surroundings of full sunlight and tall grasses and reeds; when bred in zoos which do not provide this type of habitat, tigers are still striped.

SAQ 33 Explain how the blood of llamas provides an example of this type of adaptation.

Physiological versatility. This is the ability of the organism to function in a range of changing conditions. An example of this is provided by the plaice, which is able to change colour in a short space of time to match its background.

SAQ 34 Explain how the homeostatic regulation of body temperature in a mammal provides an example of this type of adaptation.

The organism is able to survive in a wide range of conditions because it can make appropriate and rapid adjustments to changes in these conditions.

All organisms are physiologically tolerant to a certain extent. Perhaps the least flexible are those living in the most constant environments. The abyssal depths of the oceans are among the least changeable of all the Earth's habitats, and it is perhaps not surprising to find animals there which display some of the most extraordinary adaptations.

SAQ 35 Why do you suppose humans are a more successful species than the giant panda when dietary tolerance is considered?

SAQ 36 The following are characteristics of many deep sea fish. Can you relate each of the features to the conditions in the sea at depth?
(*a*) Bioluminescence
(*b*) No swim bladder (an air-filled sac which enables bony fish to maintain neutral buoyancy)
(*c*) Active at low temperatures

Developmental flexibility. This type of adaptation is said to occur when the organism is transferred into different conditions, and a change in structure occurs which better suits the organism's new circumstances. A good example of this is seen in goldfish adapted to saline waters. These fish change their kidney structure and functioning to nearer that of marine fish.

SAQ 37 Explain how the changes in the blood of humans in response to longer-term high altitude shows developmental flexibility.

It is worth remembering, however, that not all changes are advantageous. Some may be an indication that the organism cannot cope with the pressure of the environment.

SAQ 38 Explain how primates, with their highly developed ability to learn from experience, are at an advantage over animals with less sophisticated brains in the face of changing environmental conditions.

4.2.1 Related adaptations

Adaptations take every form: structural (such as birds' feet may be adapted for swimming, perching, killing prey), functional (such as countercurrent heat-conserving mechanisms in dolphins' flippers), biochemical (such as specificity of enzymes) and behavioural (such as courtship and migration).

All adaptations to a given way of life are relative to each other. For example, the behavioural heat reduction adaptations in the camel (considered in section 3.3.1) are related to a host of other adaptations of the camel to its desert life, for example, heavy coat, large flat feet and physiological mechanisms for water conservation. Groups of related adaptations like this are called **adaptive suites.**

SAQ 39 List the related adaptations for diving seen in the seal.

4.2.2 Niches

The important characteristics of any group of organisms are those which enable it to take full advantage of as many opportunities as present themselves. One aspect of the success of a group could perhaps be measured by the number of niches it can exploit. A **niche** can be defined as the sum total of all the biological activities of a species. These include habitat, social interaction, predation, disease, parasitism and the effects of competition.

Niches have also been described as 'ecological opportunities in time and space'.

Competition for niches occurs amongst animals and amongst plants. The best-adapted organisms successfully occupy and exploit them, and reproduce successfully under the conditions of the niches in the succeeding generations.

A good example of this is seen in the peppered moth, *Biston betularia*, which exists in two forms, dark (melanistic) and light (figure 18). There is greater selection pressure from predation by birds on the light form of *B. betularia* in the industrial north of the country, so the dark form predominates there. For similar reasons, the pale moth is seen more frequently in the south west.

18 White and black (melanistic) specimens of *Biston betularia* on lichen-covered bark. Left: white, right: black

In order to get the best out of a niche, an organism must be a 'specialised generalist'. The number of different adaptations on the basic plan of the organism, be it a flowering plant, a mammal or insect, reflects the various niches that a group has been able to occupy. This is why there are so many variations on the basic patterns of organisms, why there is such a rich variety of life.

SAQ 40 Two genera of waterboatman, *Notonecta* and *Corixa* live in freshwater. *Notonecta* is a fierce predator, while *Corixa* is a scavenger, feeding on dead and decaying plant and animal material. Explain why, although these insects both occupy the same habitat, they occupy different niches.

4.3 Descent with modification

As we saw in section 1, being adapted is key to the long-term survival of any group of organisms. So any theory of evolution must explain long-term changes and provide a mechanism to account for them. Modern evolutionary theory, as we shall see in section 5, can be seen to consist of two main theories or groups of theories. There are those dealing with natural selection, the mechanism of evolution, which is dealt with later in section 5. However, the theory of natural selection depends upon the prior theory that there is *descent with modification*, that is that each species arose from a preceding species that can be traced back to one primal type. The theory of descent with modification can be seen as the concept of evolution itself and applies to living and fossil organisms, but it is not a theory about cosmic or cultural evolution.

Nowadays the descent theory is the basis of modern classification systems. In other words, the theory of natural selection could, in principle, be modified or discarded without changing our fundamental view of the origin and relationship of present-day organisms and fossils. So what then is the evidence for descent with modification? Broadly speaking, it is the fossil record.

4.4 Fossil evidence

4.4.1 What is a fossil?

Fossils are any sort of evidence of once-living organisms. The study of fossils is called **palaeontology**. Parts or whole bodies of animals and plants, or signs (such as footmarks) of their presence can be preserved for millions of years. Fossils can be preserved in rock, ice, peat and amber (the sticky resin of some conifers). It is quite rare for whole organisms, or even substantial parts of them, to be preserved. Commonly, the soft parts decay or are eaten before fossilisation can take place. The hard parts such as bones, teeth and shells are usually all that is preserved.

Fossils were not always understood to be the remains of dead organisms; for instance, the Aristotelians construed fossils as natural growth from the sperm of animals that fell onto rocks, just as living organisms were natural growths from sperm that fell onto or into living matter. Understanding of the relationship between fossils and organisms only began to develop in the seventeenth century. Scientific appreciation of the fossil record started in the eighteenth century when, paradoxically, its interpretation by the French scientist Cuvier was held to be evidence against evolution!

Modern understanding of fossils has depended on advances in other fields of science and now many of the problems and interpretations of the past (and the tensions with religion) are forgotten. Arguably it is the way evolutionary theory fits with (is consilient with) other scientific theories that lends it such strength.

4.4.2 The fossil record

How complete a picture of the fossil record do we have?

Firstly, inevitably, it is fragmentary. Fossils are laid down in sedimentary rocks. So we can only glean data from those groups of organisms that died under conditions favouring fossilisation and which had suitable hard parts. Some of our best and longest records are therefore, not surprisingly, of marine organisms such as foraminifera and trilobites. The richness of the fossils and the length of the record may vary from one site to another; the Grand Canyon has a particularly long record. The Grand Canyon is composed of many layers of sedimentary rock. The number and diversity of fossils becomes less with increasing depth. Fossil reptiles, for example, appear only in the upper (more recent) layers, while fossil annelids appear further down. This decrease in number and variety is typical of the fossil record.

Secondly, the fossil record cannot always be dated accurately. Sometimes the fossil itself can be dated from the carbon-14 radioactivity levels, and sometimes the source deposit can be dated, for example, using potassium–argon, strontium, rubidium and lead dating techniques. On other occasions it may be necessary to rely on correlating the source deposit with another deposit of a known age.

Thirdly, our understanding of the fossil record may be modified by evidence from other sources. For many years palaeontologists had suggested that chimps and man diverged from a common ancestor about 30 million years (Myr) ago. More recently immunological evidence based on study of blood albumen, together with DNA binding studies, have forced palaeontologists to revise their estimates to between 4 and 8 Myr.

Clearly for these reasons and others we can never expect to write a definitive history of life on Earth, there will always be gaps and areas open to interpretation.

SAQ 41 If, for example, the earliest land animal is dated at x Myr ago, this does not mean that animals colonised land at this time, nor does it mean that it was first done by animals of this type. Explain why not.

Nevertheless the evidence that we do have fits into a coherent picture and we are able to at least outline the evolution of the major groups. Figure 19 gives the major geological eras together with certain evolutionary landmarks. Certain parts of the fossil record are so interesting or so complete that they merit further study.

4.4.3 *Archaeopteryx*

Sometimes the fossil record reveals forms which are intermediate between two present-day types, often called 'missing links'. These discoveries are always greeted with much interest and a great deal of information about the forerunners of the present-day species can often be pieced together.

The discovery of *Archaeopteryx*, a fossil form intermediate between the reptiles and birds, pointed to the possibility that these two groups share a common ancestor (see figure 20). In fact such a fossil had been predicted in principle only two years before the discovery of the first specimen. *Archaeopteryx* had teeth and a long tail (reptile characteristics) but it also had feathers and wing-like forelimbs (bird characteristics).

19 The history of life as revealed by the fossil record

Eras	*Periods*	*Epochs*	*Aquatic life*	*Terrestrial life*
with approximate starting dates in millions of years ago				
			Periodic glaciation and arid climate	
	Permian 280 ± 10		Extinction of trilobites, placoderms	Reptiles abundant (cotylosaurs, pelycosaurs). Cycads and conifers; ginkgos
			Warm humid climate	
	Pennsylvanian 310 ± 10	**Carboniferous**	Ammonites, bony fishes	**Age of amphibia** First insects, centipedes First bryophytes. First reptiles Coal swamps
			Warm humid climate	
	Mississippian 345 ± 10	**Carboniferous** / **Age of fishes**	Adaptive radiation of sharks	**Age of amphibia** Forests of lycopsids, sphenopsids, and seed ferns. Amphibians abundant. Land snails
			Periodic aridity	
Palaeozoic 600 ± 50	**Devonian** 405 ± 10		Placoderms, cartilaginous and bony fishes Ammonites, nautiloids	Forests of lycopsids and sphenopsids. Ferns. First gymnosperms Millipedes, spiders First amphibians
			Extensive inland seas	
	Silurian 425 ± 10		Adaptive radiation of ostracoderms; eurypterids	First land plants (psilopsids, lycopsids) Arachnids (scorpions)
			Mild climate, inland seas	
	Ordovician 500 ± 10		First vertebrates (ostracoderms) Nautiloids, *Pilina*, other molluscs Trilobites abundant	none
			Mild climate, inland seas	
	Cambrian 600 ± 50		Trilobites dominant First eurypterids, crustaceans Molluscs, echinoderms Sponges, cnidarians, annelids Tunicates	none
			Periodic glaciation	
Pre-Cambrian	2000		Fossils rare but many protistan and invertebrate phyla probably present	none

Continued overleaf

Eras	Periods	Epochs	Aquatic life		Terrestrial life
with approximate starting dates in millions of years ago					
	Quaternary 0.5–3	**Recent**		Age of mammals	Humans in the New World
		Pleistocene	*Periodic glaciation*		First humans
Cainozoic 63 ± 2	**Tertiary** 63 ± 2	**Pliocene**	All modern groups present		Hominids and pongids
		Miocene			Monkeys and ancestors of apes
		Oligocene			Adaptive radiation of birds
		Eocene			Modern mammals and herbaceous angiosperms
		Palaeocene			
Mesozoic 230 ± 10	**Cretaceous** 135 ± 5	Age of reptiles	*Mountain building (e.g. Rockies, Andes) at end of period*		
			Modern bony fishes Extinction of ammonites, plesiosaurs, ichthyosours		Extinction of dinosaurs, pterosaurs Rise of woody angiosperms, snakes
	Jurassic 180 ± 5		*Inland seas*		
			Plesiosaurs, ichthyosaurs abundant Ammonites again abundant Skates, rays, and bony fishes abundant		Dinosaurs dominant First lizards. *Archaeopteryx* Insects abundant First mammals First angiosperms
	Triassic 230 ± 10		*Warm climate, many deserts*		
			First plesiosaurs, ichthyosaurs Ammonites abundant at first Rise of bony fishes		Adaptive radiation of reptiles (thecodonts, therapsids, turtles, crocodiles, first dinosaurs, rhynchocephalians)

20 Cast fossil of *Archaeophteryx*

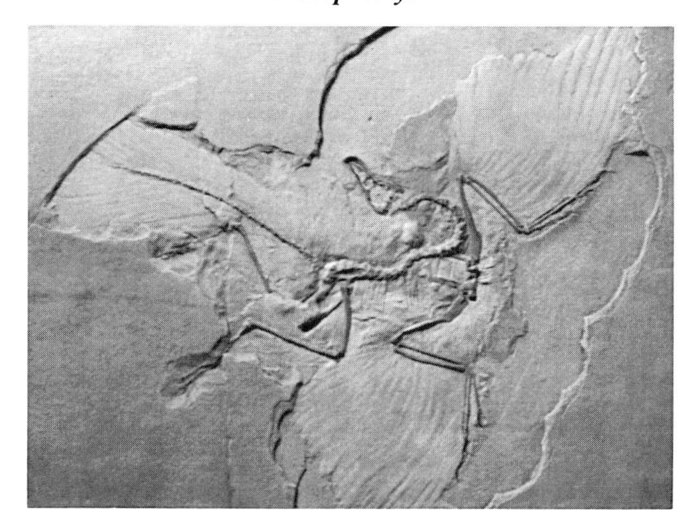

These so-called 'missing links' are very rare but provide strong evidence for evolution. Another example is the caterpillar-like creature *Peripatus* which has a variety of arthropod and annelid characteristics.

4.4.4 The horse

One of the most complete examples of fossil record is that of the horse. Evidence has been unearthed which illustrates the development of the horse over the last 60 Myr.

The modern horse is superbly adapted to its natural lifestyle, that of grazing and rapid running. It has long slender legs, with single hooves, large skeletal muscles, an elongated muzzle and high-crowned, open-rooted molars and premolars. These adaptations can be seen today in the wild horses and zebras. The domestic horse has been especially bred for such characteristics as strength and speed. All these forms arose from a common ancestor which appeared in the Eocene period. Strictly this ancestor should be called *Hyracotherium* but it is commonly referred to as 'Eohippus'.

SAQ 42 When was the Eocene period?

'Eohippus' was less than 30 cm at the shoulder. In many respects it probably resembled a small horse but it had more than one digit on each foot. Figure 21 shows that the third digit (equivalent to our middle finger) was becoming dominant in 'Eohippus'.

'Eohippus' lived in wet forest land, so the four digits on its front feet and the three on its hind feet may have helped to spread its weight and thus prevent it from sinking into the soft ground. Some of the major features of the development of the horse from this relatively unspecialised ancestor can be traced in figure 21. Two main types that came from 'Eohippus' were *Miohippus* and *Merychippus*. Unlike *Miohippus, Merychippus* became adapted to life on the plains, becoming a grazer instead of a browser. This change was correlated with the spread of savannah environments in North America during the Oligocene and Miocene, followed by a change to arid steppe conditions in the Pliocene and Pleistocene and the consequent increase in the area of grassland. Furthermore, there were probably few animals of a comparable size to compete with *Merychippus* during this evolutionary change. *Merychippus* became adapted for grazing with an elongated muzzle (the typical herbivore diastema) and ridges of hard enamel on its teeth for grinding the tougher grass leaves.

The major trends in the evolution of the horse (changes in cranial, dental and postcranial structures, feet and size) had been established by the beginning of this century. However the detailed interpretation of this history is still a matter of debate. Huxley and others in the early twentieth century had interpreted the fossil evidence as a single evolutionary trend, accounted for by 'orthogenesis'. The theory of **orthogenesis** (inherent trends due to the genetic material) has now been discredited, and anyway by the mid-twentieth century it was clear that the phyletic pattern of the horses (Equidae) was too complex to be accounted for by orthogenesis.

Nowadays the phyletic history is seen as very complex, the origin of several of the genera has been put back and it seems likely that there have been at least three major intercontinental dispersals. Figure 22 shows a modern interpretation.

21 Table showing the evolution of skull, feet and teeth of the horse

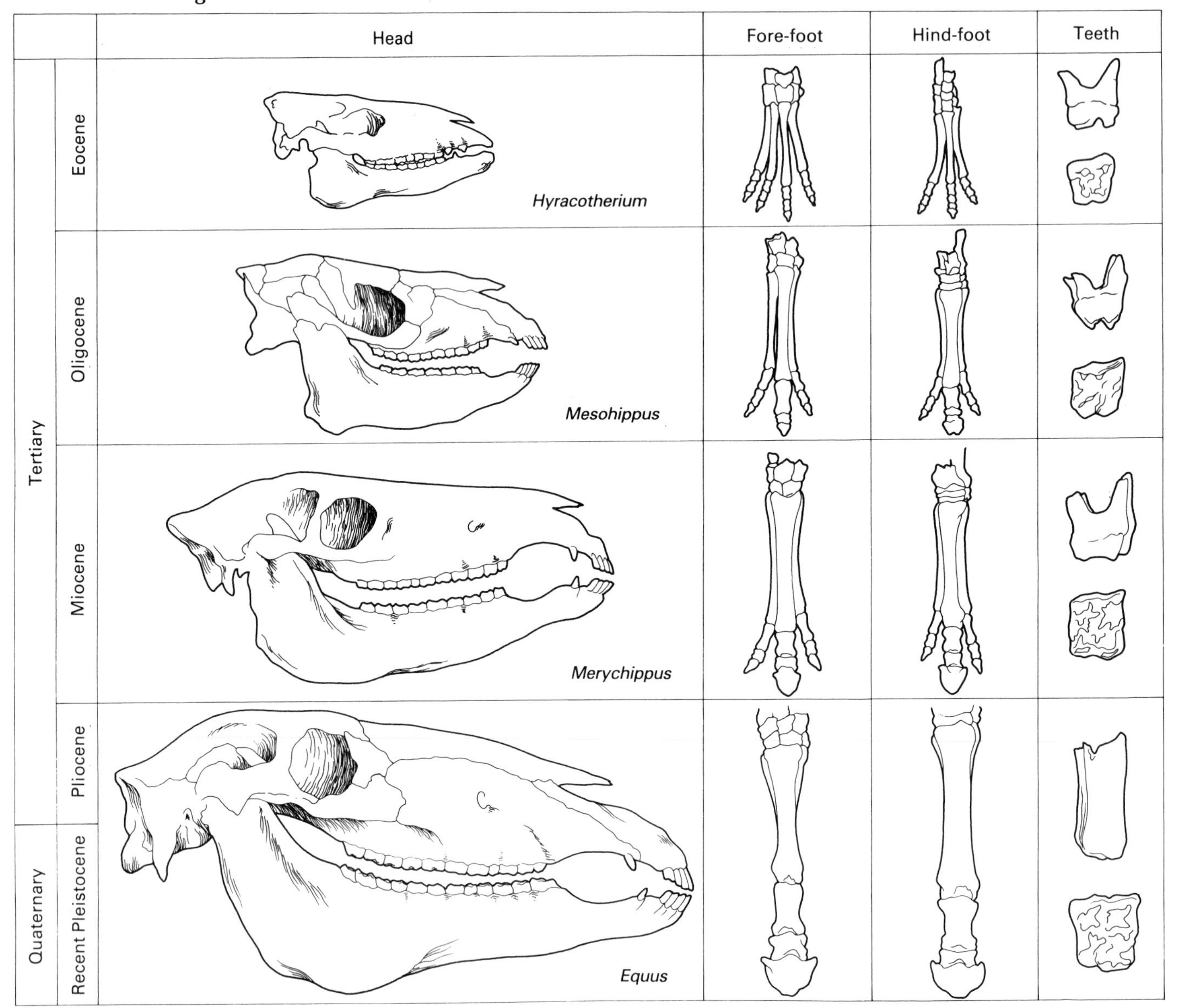

SAQ 43 Place the following in evolutionary order: (*a*) *Dinohippus*, (*b*) *Merychippus*, (*c*) *Hyracotherium*, (*d*) *Miohippus*, (*e*) *Equus*.

SAQ 44 What was the common ancestor of *Equus* and *Nannipus?*

SAQ 45 When did the last browsing horses appear to die out?

SAQ 46 What happened to the population of *Equus* in North and South America in the late Quaternary?

SAQ 47 Can you make any suggestions to account for this?

SAQ 48 Where do present-day horses in America come from?

Furthermore the general trends observed in the evolution towards *Equus* should not be thought of as inevitable or universal in the Equidae. For instance *Nannipus* was significantly smaller than its ancestor *Merychippus*. Many other features show fluctuations or reversals at different times and in different places.

22 Current phylogeny of the Equidae

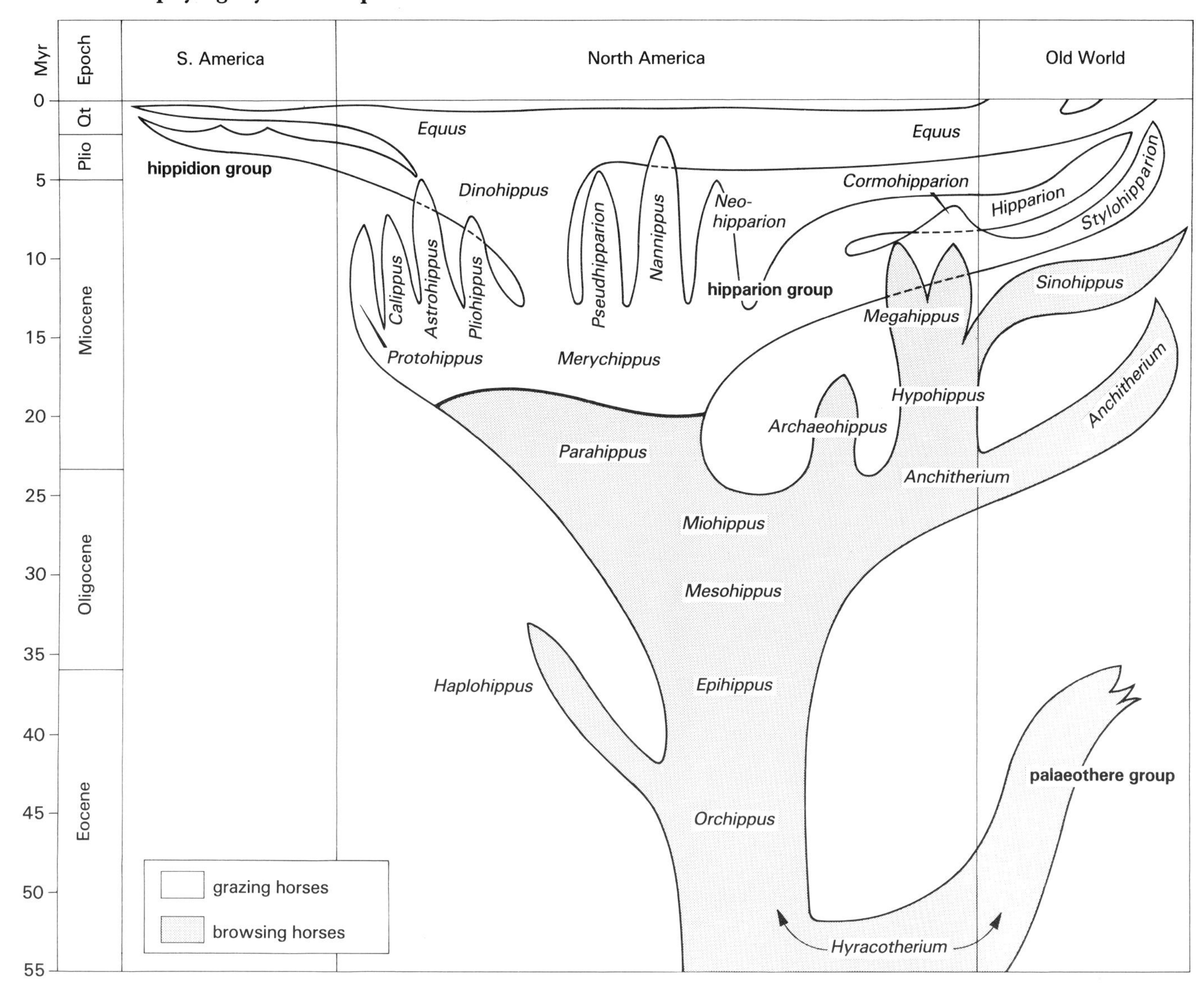

You will have noticed by now that our concept of adaptation has enabled us to make sense of the record. We can see how the main evolutionary changes are likely to improve fitness in a world where forest is retreating and grassland expanding. Our view of adaptation also enables us to make a wider speculative reconstruction of horse evolution (thanks to the wealth of fossil remains). For instance, some Miocene remains show distinct age classes which allows us to infer seasonal breeding patterns; different species show different longevities; it is also possible to make hypotheses about the intensity of natural selection at different times. Analysis of the facial fossae and comparison with modern equids suggests that even *Hyracotherium* lived in vocal social groups.

Although we are able to build a quite detailed picture of the macroevolutionary phenomena of equid phylogeny, the fossil record does not allow us to examine the detail of microevolution. Horse fossil evidence, in other words, supports the theory of descent with modification, but does not allow us to properly test either the theory of natural selection or modern challenges. The mechanism of evolution is dealt with in section 5.

4.5 Origins, replacements and extinctions

4.5.1 The origin of the fossil record

The Earth is reckoned to be about 4.5 billion years old which places its origin in a very early geochronological aeon (see figure 23). Although we have no terrestrial rocks from this, the Hadean, aeon in which to search, it seems that life originated in the following aeon, the Archean. Assigning a likely date to the origin of life is a difficult task for several reasons. Two of these are, firstly, it is necessary to decide what constitutes life rather than prebiotic chemistry. Secondly, the interpretation of the nature of the earliest organic carbon compounds is problematic. Earliest life, microbial and largely anaerobic, is considered to belong to the Archean aeon, but the earliest widespread fossils scientists can identify are probably the stromatolites of the Proterozoic aeon, 3.5 to 0.6 billion years ago (see figure 24). Stromatolites are interpreted as trace fossils, primarily of photosynthetic microbial

23 Geochronological aeons

Aeon		*Billions of years BP*		Major events
Chaotic	'PreCambrian' (Chaotic to Proterozoic)	6.0(?)–4.6*	No dated terrestrial rocks (Chaotic to Hadean)	Origin of solar system; first major thermal events; no direct age measurements
Hadean		4.6–3.9		Establishment of Earth–Moon system; formation and differentiation of Earth as a planet (core, crust, atmosphere); oceans formed by about 4 billion years ago; lunar rocks from Apollo 12, 14, 15 dated at 4.5 billion years, from Apollo 11 at 3.6 billion
Archean		3.9–2.6	Continuous record of dated terrestrial rocks (Archean to Phanerozoic)	Tectonism; protocontinental rocks (Greenland); geology 'immature'; earliest microbial life, age of anaerobes
Proterozoic		2.6–0.6		Modern geological tectonic and weathering patterns; stromatolite communities; cyanobacteria
Phanerozoic	Classical fossil record	0.6–0		Animals and plants; reef communities

*Cosmologists consider the universe to be less than 20 billion years old

communities. Peculiar microfossils of the Proterozoic aeon suggest that eukaryotes had evolved by one billion years before present (BP).

24 Living stromatolites at Shark Bay, Australia

There is some debate about how eukaryotic cells arose. One theory is that they originated by a series of symbiotic unions of various prokaryotes, thus giving rise to organelles such as mitochondria and chloroplasts. However it is hard to recognise the earliest eukaryotes because the cellular ultrastructure is rapidly destroyed when the cell dies (usually by its own enzymes) and so the subcellular organisation is not fossilised.

The dramatic expansion of animal life began in the Phanerozoic aeon (the present aeon which began about 600 million years ago) and may have been due to the availability of significant amounts of free atmospheric oxygen. Those organisms that were able to use this for aerobic respiration had numerous niches available for exploitation. The first era of the Phanerozoic, the Palaeozoic, is characterised by a decline in stromatolites, a rise in calcium carbonate reefs and the world-wide appearance of skeletalised metazoans. This marks the beginning of the classical fossil record, as outlined in figure 19.

4.5.2 Replacements and extinctions

The classical fossil record reveals the phylogeny or evolutionary history of the main groups of animals. But it should be remembered that this phylogeny has not occurred against a stable background. During the Phanerozoic the magnetic poles of the Earth have shifted, there has been substantial continental drift and reshaping of land masses leading to considerable variations in climate in different geographical locations at different times. Continental drift has also had a profound effect on the dispersal of organisms, particularly land animals.

Nor does the fossil record reveal a story of steady evolution. Different eras are characterised by the emergence or dominance of a particular group or groups of organisms. For instance, in the late Ordovician (450 Myr BP) the first fossil vertebrates appeared, jawless fish. Fish then became abundant and dominant. Some of the present-day types, cartilaginous fish, bony fish and lung fish, appeared in the Silurian, 400 Myr BP.

During the Permian (250 Myr BP) great climatic and geographical changes took place. As well as more variation in climate and a great glaciation that spread across the southern hemisphere, there was a general upheaval and folding of the Earth's crust which produced great mountain ranges. These changes provided a number of new terrestrial environments. For those organisms that could survive, swamps, plains, deserts and mountains became available. Great structural and functional changes in organisms took place. It was at this time that a transition to terrestrial life occurred.

SAQ 49 Why is the Carboniferous period (about 300 Myr ago) so called? What characterised it?

During the Mesozoic era there were great changes in the plant kingdom. The great bryophytes of the Carboniferous were replaced by much smaller ferns, lycopods and horsetails. One present-day genus which originated at about that time is *Ginkgo*, possibly the oldest living seed-bearing plant (figure 25). The flowering plants (angiosperms) arose about 100Myr ago (see figure 26).

SAQ 50 Suggest what structural changes may be associated with this succession of plant types.

The Mesozoic is sometimes known as the 'age of reptiles' because of their dominance in all the major

25 A fossilised *Ginkgo* leaf

environments. Fish, amphibia and reptiles dominated for over 125 Myr and, during this time, the mammals and birds made their separate appearances.

Clearly the evolutionary process is not simply a matter of new groups emerging and adding to the sum of species, families, phyla and so on. Equally important is the process of extinction. There are various mechanisms that may give rise to extinction and different scales of extinction. We shall consider some of these.

The term extinction usually evokes the idea of a single catastrophic event; yet this can be far from the case. When one group evolves into another (such as *Mesohippus* to *Miohippus* in figure 22) then we have a **taxonomic extinction**, yet there is as much continuity as if the form of the descendants had not

26 History of the plant kingdom

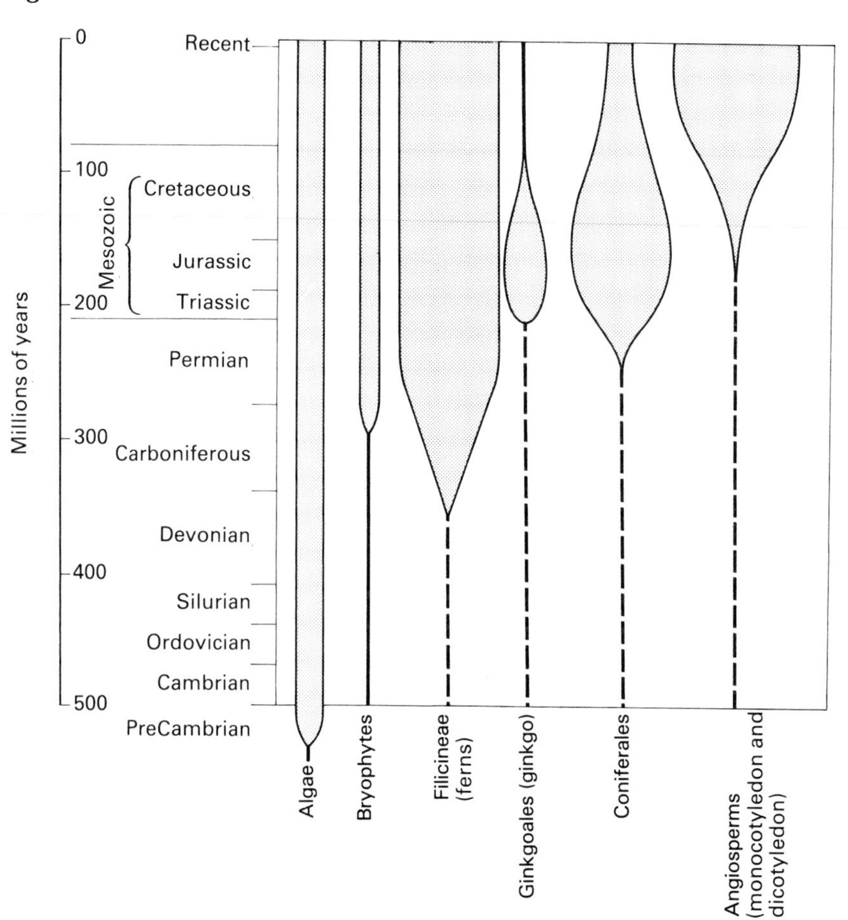

changed. In the fossil record the disappearance of the earlier form may occur at different times in different places giving the illusion of multiple extinctions.

Local extinctions sometimes occur, but as long as reinvasion remains a possibility such local extinctions are repeatable. However, if the death rate of a species exceeds the birth rate for long enough then this may result in extinction for ever, sometimes called **blanket, total** or **global extinction**. This is obviously far more serious than local extinction which is really a reduction in the range of a species.

Coextinctions occur when the extinction of one taxon brings about the extinction of another dependent taxon. Animal–plant extinction may occur if a plant species (or sometimes a plant community) disappears. Commensals, scavengers and parasites can all be expected to disappear if their host becomes extinct. Such extinctions are logical extensions of ecological principles you have studied in unit 9, *Ecology*, and are discussed further in section 6 of this unit. Most bird extinctions in North America are attributed to the disappearance of the mammal megafauna in the Pleistocene and the subsequent loss of niches for bird scavengers and commensals. During the Pleistocene the mastodont became extinct and consequently its predator *Smilodon* (figure 27).

27 Mastodont and *Smilodon*

The cause of extinction is usually a failure of a species to adapt to a change in the environment, such as intensifying seasonality, or a secondary consequence of this, such as change in vegetation. Sometimes there appear to be extinction episodes when extinction rates exceed origination rates, leading to decreased taxonomic diversity. Fortunately the reverse occurs during other episodes. One mechanism leading to rapid taxonomic diversification is adaptive radiation.

4.6 Adaptive radiation

4.6.1 Basic principles

Occasionally, a breakthrough occurs in the design of organisms (such as their body plan) and this allows a burst of evolution to occur as new species are able to spread out, diversify and colonise new niches. This is known as **adaptive radiation** and such phenomena are one of the reasons why evolution is not a steady progression.

One of the greatest animal adaptive radiations occurred as a result of the conquest of land by vertebrates, some 400 to 300 Myr ago.

SAQ 51 What is the succession of organisms which colonised the land likely to have been?

The first vertebrates to have lived on land were amphibian-like organisms called labyrinthodonts (figure 28). They had primitive lungs which could

28 Labyrinthodont

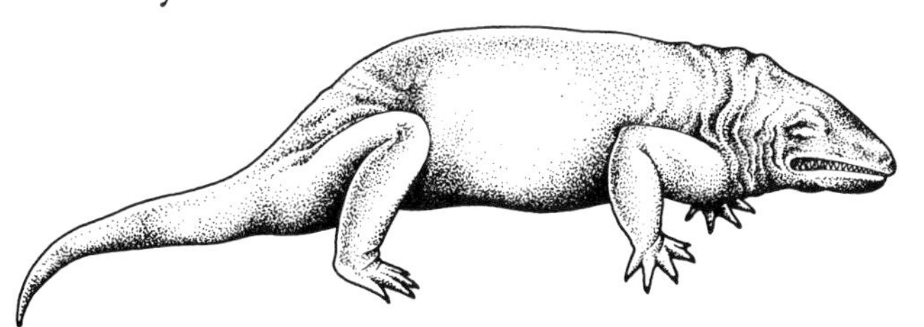

breathe air and paired fins which they used to pull themselves along. However, in all other respects they more closely resembled fish than the modern amphibians.

SAQ 52 What are the two main problems facing animals when they change from aquatic to terrestrial life?

Another great adaptive radiation which produced an expansion of the number of species occurred 300 to 250 Myr ago. This was during the so-called 'age of reptiles', when, amongst other groups, the great dinosaurs emerged. The reptiles became dominant over the amphibians because they were independent of water for the purpose of reproduction.

SAQ 53 What was the special adaptation that enabled the reptiles to reproduce away from water?

About 60 Myr ago, another great adaptive radiation among the vertebrates occurred. This was that of the mammals and produced an enormous diversity of form including bats, carnivores, rodents, ungulates, whales and primates.

4.6.2 Adaptive radiation in mammals

The first mammals appeared in the Jurassic period of the Mesozoic (180 to 135 Myr ago). Yet it was not until the Cainozoic era (about 100 Myr later) that they began to achieve dominance over the reptiles. There appear to have been several factors which contributed to this history.

Firstly what was the origin of the mammals? It is debatable whether we should define the mammals as deriving from a single ancestor **(monophyletic)** or from several ancestors **(polyphyletic)**. It is a truism that if one goes back far enough one will find a common ancestor (this is true for any groups of any organism). What is at issue here is whether the common ancestor that one reaches is a mammal or a reptile. The trouble is that the reptile–mammal transition is gradual and the current argument amongst palaeontologists seems to hinge on details of the articulation of the jaw.

Modern mammals are clearly defined by their morphological and physiological innovations such as having a single (dentary) jaw bone, homeothermy,

29 *Cynognathus*, the 'dog-jawed' reptile, a therapsid, possibly an early mammal

hair, live birth and suckling of the young, a diaphragm, an enlarged brain and four-chambered heart. But some of these features may have been shared with the then mammal-like reptiles, which later became extinct leaving a more clear-cut mammal–reptile distinction.

A resolution of the history of the origin of the mammals is not necessary to an appreciation of the adaptive radiation that occurred in the Cainozoic. The main points are as follows.

By the Cainozoic, mammals were widespread but small and comparatively inconspicuous, superficially not unlike modern shrews and hedgehogs. At the end of the Cretaceous the dinosaurs became extinct. Debate about the cause of this is still continuing, but most palaeontologists nowadays agree that it seems to have been a gradual process and occurred at different rates in different geographical locations. Whatever the reason, the demise of the dinosaurs left many niches available for the mammals to exploit. In the absence of the dinosaurs even quite unspecialised mammals were able to exploit the new resources, but as their niches became 'filled up' intra- and interspecific competition would have led to increased specialisation and more obvious adaptation.

Various examples include the earliest bats which appear to have evolved quickly to exploit the niches vacated by the flying reptiles. The evolution of large plant-eating mammals to replace the herbivorous dinosaurs is particularly characteristic of the early Cainozoic. Many of the varieties of early mammal will have undergone their own adaptive radiations; we have seen one particular example of this already, the horse.

30 Eutherian mammals with metatherian counterparts

Adaptive radiation of the mammals began between 70 and 60 Myr ago by which time two major groups had emerged, the pouched mammals, marsupials or **Metatheria** and the placental mammals or **Eutheria**. It is preferable to use the terms Metatheria and Eutheria because it is now known that marsupial embryos have a rudimentary placenta. The third group of mammals, the egg-laying monotremes, have only three modern descendant species and their phyletic relationship to the Metatheria and Eutheria is still a point of contention.

As you can see from figure 30, the various forms which have appeared in the Metatheria have eutherian counterparts.

SAQ 54 Make a two-column table. Enter the generic name of each metatherian in one column and in the other put the name of its eutherian counterpart.

SAQ 55 How do you account for the similarity of form between groups of mammals whose common ancestors were different?

Present-day orders of eutherian mammals have

31 Diagram showing probable radiation of present-day groups from early mammal types

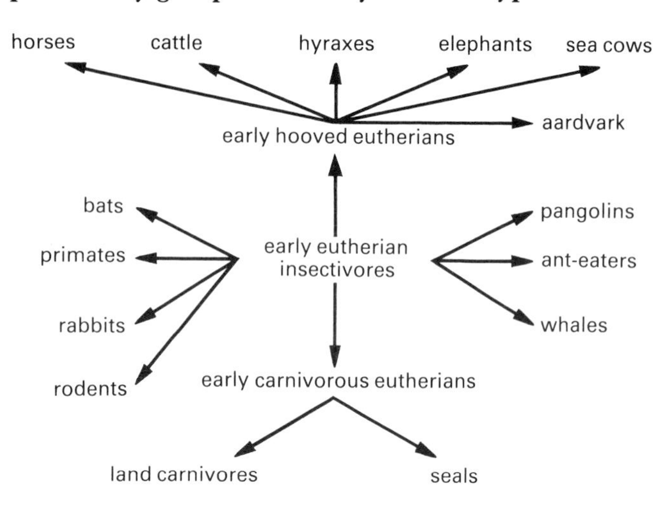

probably radiated from an early, relatively unspecialised, insectivorous ancestor (see figure 31). The divergence in form from the original was due to the selection pressures operating on them in the various niches which they came to occupy. Today's tree shrew is probably the nearest living eutherian to that ancestral insectivore (figure 32). The tree shrews have changed least over the course of the last 60 Myr and are the most primitive eutherians alive today. From this mammalian ancestor have risen the eutherian orders of today (see figure 33).

32 Tree shrew

33 Eutherian orders

Order		*Families*	*Species*
Edentates	sloths, armadilloes, American anteaters	3	29
Insectivora	shrews, moles, tenrecs etc.	6	343
Scandentia	tree shrews	1	16
Dermoptera	flying 'lemurs'	1	2
Chiroptera	bats	19	950
Primates	lemurs, lorises, monkeys, apes, man	11	179
Carnivora	dogs, bears, cats etc.	8	240
Pinnipedia	seals, sea-lions	3	34
Cetacea	whales, dolphins	9	76
Sirenia	sea cows	2	5
Proboscidea	elephants	1	2
Perissodactyla	odd-toed ungulates	3	17
Hyracoidea	hyraxes	1	5
Tubulidentata	aardvark	1	1
Artiodactyla	even-toed ungulates	10	184
Pholidota	pangolins	1	7
Rodentia	rodents	20	1581
Lagomorpha	rabbits, hares, pika	2	54

SAQ 56 List the orders that have fewer than 10 species.

These poorly represented orders are known as **relict orders.** The fact that only six orders have more than 100 species may indicate that mammals are in a decline. The six most successful orders are:

Artiodactyla	e.g. pig
Primate	e.g. orang utan
Carnivora	e.g. tiger
Insectivora	e.g. hedgehog
Chiroptera	e.g. fruit bat
Rodentia	e.g. mouse

34 Representatives of the most successful mammal orders

(a) Pig

(b) Hedgehog

(c) Tiger

(d) Orang utan

(e) Fruit bat

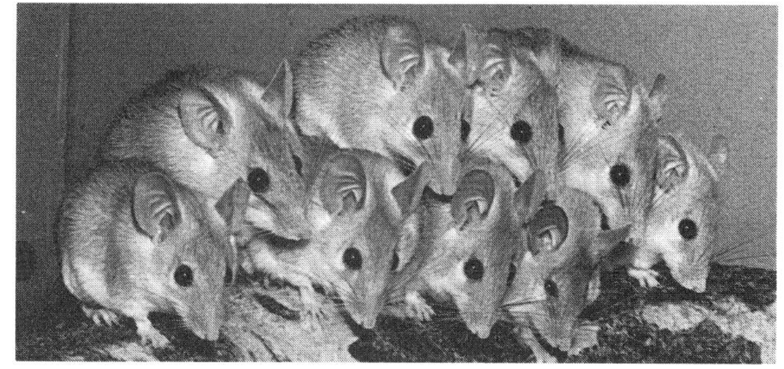
(f) Mouse

SAQ 57 Make a table to summarise some of the modifications for feeding, of the basic mammalian plan, typically shown by these orders.

4.7 References and further reading

Life on Earth by D. Attenborough.
Moths, melanism and clean air by J.A. Bishop & L.M. Cook, *Sci. Am.* January 1975, **232** pp.90–9.
Wandering Land Animals by E.H. Colbert.
Adaptation by R.C. Lewontin, *Sci. Am.* **239**, pp.212–30.
The Theory of Evolution by J.M. Smith, 3rd ed.
Looking at Mammals by P. Stanbury.

Section 5 Evolution

5.1 Introduction and objectives

This section is about **evolution,** the process in which a population of organisms undergoes changes in genotype over successive generations.

After completing this section you should be able to do the following.

(*a*) Outline the evolutionary theories of Lamarck and Darwin.

(*b*) Understand the process of natural selection and quote examples of the process in action.

5.2 Evolutionary theories

Evolution is the process in which a population of organisms goes through a series of changes in genotype from generation to generation. The theory of evolution argues that the diversity of life as we see it today has arisen by modification of existing species. There has always been much speculation about the origin of present-day forms of life on Earth, and before 1820, people's ideas about evolution were based on evidence from several different sources.

(1) **Taxonomic evidence.** (See unit 1, *Inquiry and investigation in biology,* section 4.4). This shows us that there is:
(*a*) variation between individuals of a species, and
(*b*) graded differences between one species and another.

(2) **Anatomical evidence.** The idea of evolution from a common ancestor is given considerable support by the evidence of **homologous structures.** These are parts of the body which are derived from the same basic embryonic structures in different species, but do not necessarily have the same function. A good example of this is seen in the bone structure of the forelimbs of some mammals (figure 35).

(3) **Fossil evidence.** (See section 4 of this unit).

35 Homolgous structures in the forelimbs of some mammals

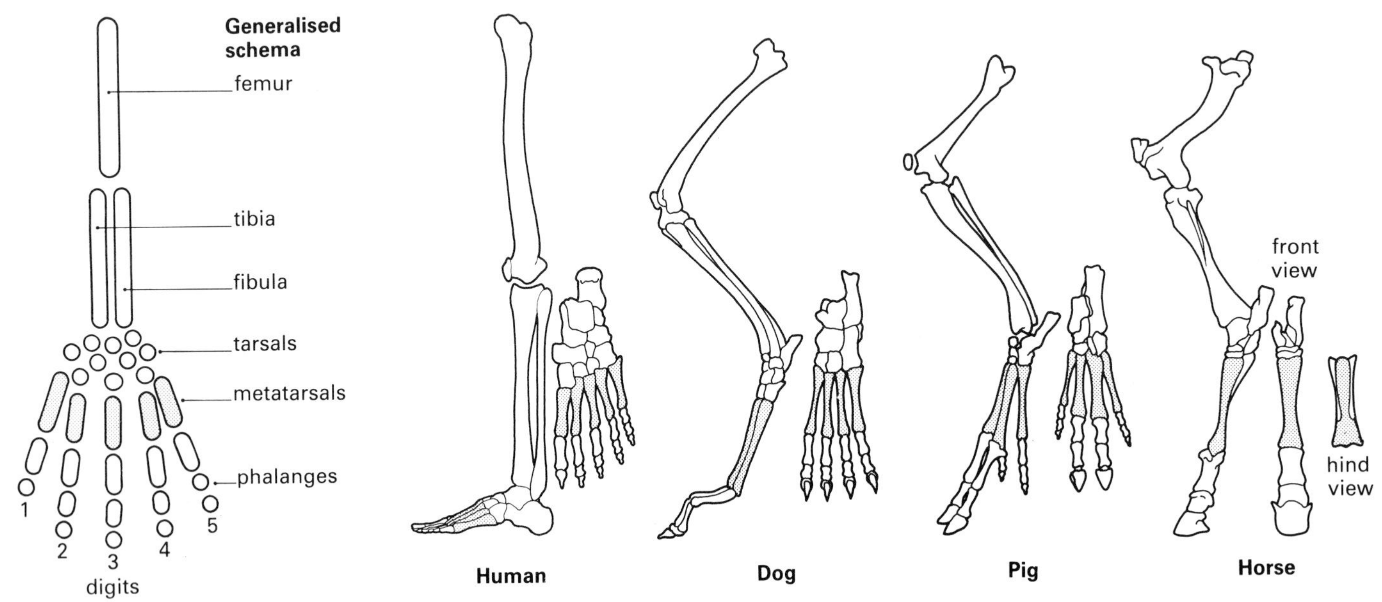

(4) **Biogeographical evidence.** There is a pattern of similarity between species occupying a geographical area. These species tend to differ from similar species found in other regions. An interesting example is seen in the island of Tristan da Cunha, which lies approximately half way between Africa and South America in the Atlantic Ocean. Of the flowering plants to be found on this island, 19 species are also found in both Africa and South America. A further seven species are found on the island and in South America only, and two more species are found on the island and Africa only (see figure 36).

36 Venn diagram showing relationship between plant species of Africa, South America and Tristan da Cunha

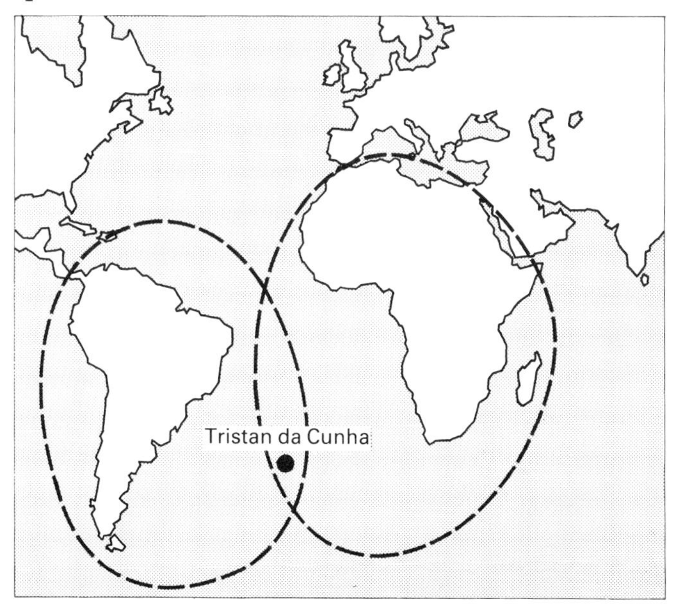

SAQ 58 Present this information as a pie chart.

The existence of related species spread along a geographical gradient provides evidence for evolution.

(5) **Selective breeding evidence.** Major changes can be brought about in a population after a few generations by manipulating the breeding. Species are not, therefore, impervious to change. This can be seen in domestic strains of plant and animals in which desirable characteristics are bred into populations of organisms, such as by gardeners and farmers.

From a wide variety of evidence, then, the evolution of one type of organism from another would seem highly probable. But what could have been the mechanism? It was against this background of evidence for evolution of organisms that biologists of the early nineteenth century sought to answer this question.

5.2.1 Lamarck

Whilst many biologists of this period accepted the evolution of organisms as a fact, the French naturalist Jean Baptiste Lamarck was the first to propose a coherent scheme as to how it came about. He suggested that organisms, within their own lifetimes, could adapt to changing environmental conditions. More importantly, these acquired characteristics could, according to Lamarck, be inherited by the offspring. Lamarck supposed that the driving force behind evolutionary change was the 'need' for organisms to perfect their adaptations. Surprisingly, it was this claim that finally led to the rejection of Lamarck's theory.

Figure 37 shows that giraffes needed to have longer necks and so stretched to perfect this adaptation.

37 Lamarckian explanation of the evolution of the giraffe's long neck

Eventually, by continual stretching, the neck lengthened and this slightly longer neck was passed on to the next generation, which carried on stretching and so on.

SAQ 59 Some fish live in underground lakes which are completely dark. Although they have eye sockets, they have no functional eyes. Using the theory of Lamarck, explain this adaptation.

Nineteenth-century biologists had no idea of the mechanism of inheritance and so Lamarck's suggestion of the inheritance of acquired characteristics was generally regarded as plausible.

5.2.2 Darwin

Charles Darwin's own observations whilst serving as naturalist on HMS Beagle (1831–6) (figure 38) had convinced him of the fact of evolution. Some specific examples impressed Darwin: island and mainland species of animals showed many similarities, for example Galapagos finches; present-day and fossil forms were sometimes remarkably similar, for example South American sloths bear a great resemblance to some fossil mammals. Darwin's key contribution to the explanation of evolutionary change was to propose a mechanism whereby it could be brought about. Darwin took an idea first presented by Thomas Malthus for human societies, and adapted it to biology. Malthus suggested that as food supplies were limited, uncontrolled reproduction would lead to competition between the people for food resources and only some would survive. Darwin saw that these conditions applied to all organisms. He then asked himself what could determine which organisms survived this struggle for existence?

Darwin noted that groups of organisms always exhibited variation, and that these variations were inherited; his own studies on domestication had shown this. Darwin proposed, therefore, that only organisms with advantageous variations would survive the struggle for existence, and, as these variations are inherited, the beneficial adaptations would be passed to offspring. He also proposed that variations arose by chance (rather than 'need' as in Lamarck's view) and therefore if these variations were beneficial they would permit the organism to

38 Journey of the HMS Beagle (1831–6)

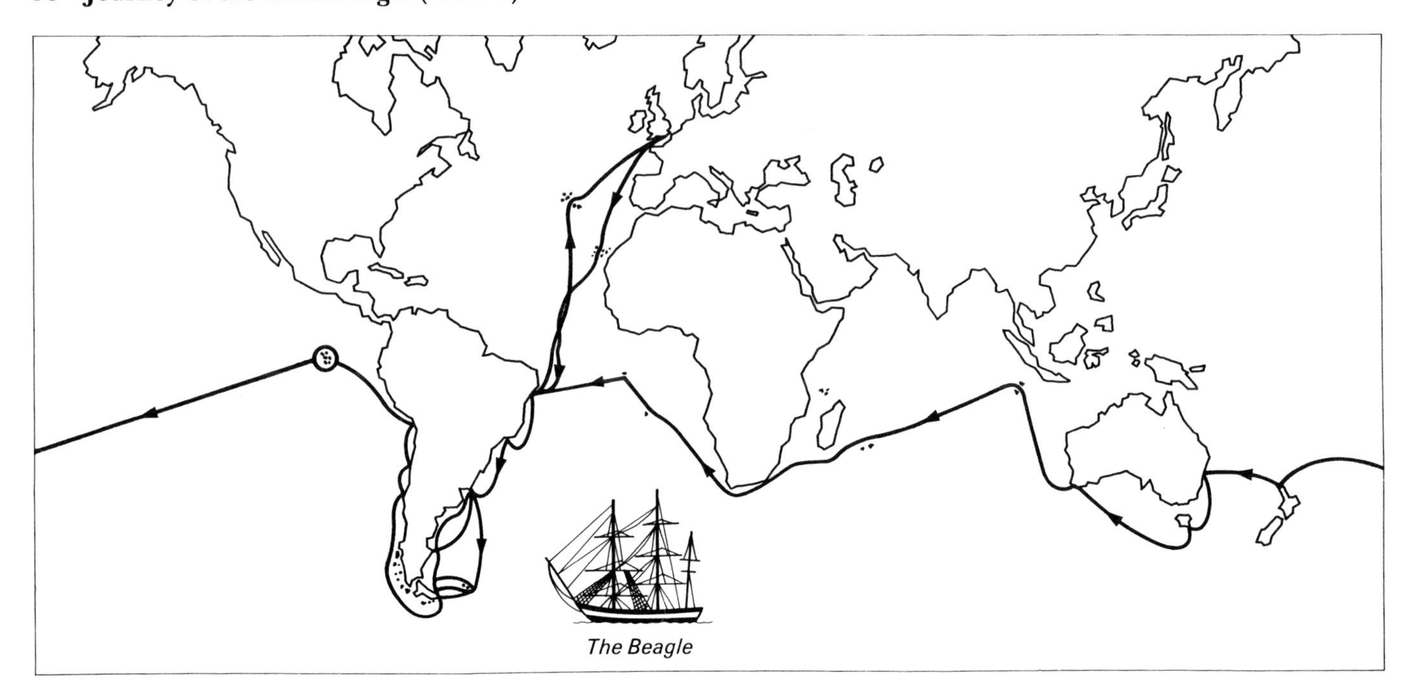

survive and reproduce, thus passing these advantageous variations to their offspring.

Independently of Darwin, Alfred Russell Wallace had produced a similar theory of natural selection.

However, in 1858 Darwin and Wallace jointly presented their theories to the Linnaean Society in London, and in 1859, Darwin published his work *On the origin of species*, in which both evidence for natural selection and evolution were presented. It caused a great controversy because it challenged the accepted views of the time.

5.3 Natural selection

Darwin's theory of natural selection offers an explanation of how change and diversity of living things can occur, and it is probably the most unifying idea in the study of living things.

It is important to realise that evolution may or may not give rise to new species. The most important fact about it is that there is a change in genotype from generation to generation. The term **speciation** (see section 5.4) is used to describe the process by which one species is transformed into two or more species over a period of time. **Natural selection** is the term Darwin used for the mechanism for evolution and speciation.

Darwin's theory of natural selection can be summarised by a series of observations and inferences.

Observation 1. Organisms have a high reproductive capability.
Observation 2. Reproductive capacity is limited by the restrictions of food supply, living space, and so on (that is selection pressures).
Inference 1. The selection pressures give rise to intraspecific competition.
Observation 3. Organisms can inherit characteristics.
Observation 4. Some characteristics are more favourable for survival (that is they are better adapted to the environment).
Inference 2. Those with the best adaptations for survival are fitter (that is they are better able to pass on their genes to the next generation).

SAQ 60 What is meant by the term intraspecific?

You should bear in mind that Darwin had no knowledge of genetic mechanisms; it was not until 1866 that Mendel's work was published, and even then, no one paid it much attention until 1900.

5.3.1 Natural selection in action

We have already looked at how selection pressures can give rise to new genotypes. The example of *Biston betularia*, the peppered moth, was cited in unit 8, *Genetics*, to show genetic polymorphism. It was H.B.D. Kettlewell, a Briton, working in the 1950s, who gathered information from the field which lead to the discovery that it was selection pressure (the predation by birds) which caused the change in gene frequencies of light and dark alleles in the industrial and rural populations of *B. betularia*. He tested this by an experiment in which he reared both light and dark moths in the laboratory, and then released both types in both environments. A summary of Kettlewell's results can be seen in figure 39.

39 Kettlewell's results: *(a)* percentages of naturally occurring light and dark moths at a rural site (Dorset wood) and an industrial site (Birmingham); *(b)* percentages of light and dark moths recaptured after release by Kettlewell

(*a*)

Area	*Light*	*Dark*
Dorset wood	95	5
Birmingham	10	90

(*b*)

Area	*Light*	*Dark*
Dorset wood	12.5	6.4
Birmingham	13.0	27.5

SAQ 61 Explain carefully what Kettlewell's results show.

You will remember that another source of variation can be due to genetic recombination and linkage. In unit 8, *Genetics*, the example quoted is the common snail, *Cepaea nemoralis*. This snail illustrates the situation of balanced polymorphism, in which two or more forms can exist in a population over several generations. Three alleles affect shell colour of *Cepaea;* they can be plain pink, plain brown or plain yellow, or have up to six different patterns of banding. The song thrush is a predator of *Cepaea.* Natural selection favours the least conspicuous snail. The environment changes as the snails move from, say, hedgerow to grass verge to leaf litter. These changes in situation cause the snails to become more or less conspicuous. There is, therefore, no 'best' colour of shell for camouflage. Several different forms are thus favoured by natural selection.

Figure 40 shows the percentage of snails visible in different habitats.

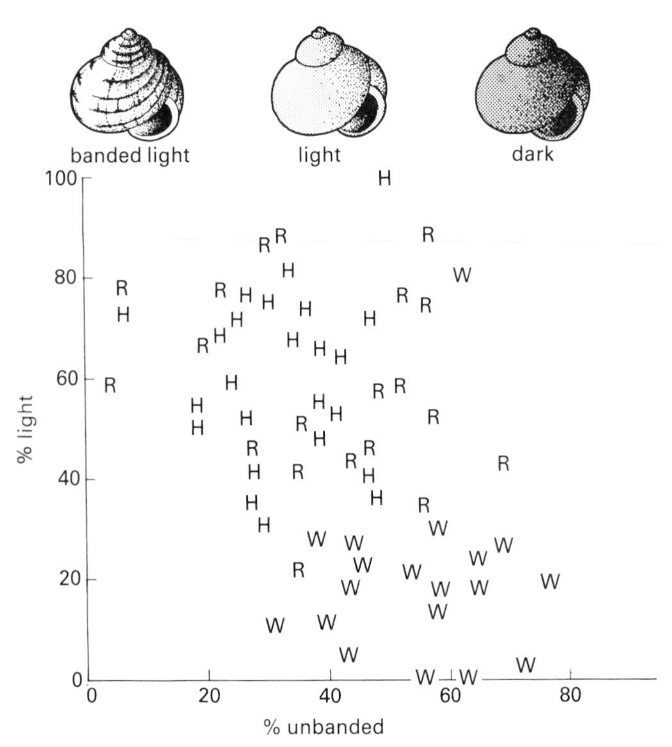

40 Scatter diagram showing changes in distribution of *Cepaea* polymorphs

SAQ 62 From the diagram, which type of shell is difficult to see (*a*) in the hedgerows, and (*b*) in the woodland?

Sickle-cell disease provides us with an example of evolution in action. Refer back to the video sequence on sickle-cell disease in unit 8, *Genetics* and to the video *Variation in populations.*

SAQ 63 Explain under what circumstances it is selectively advantageous to possess the sickle cell genotype.

5.3.2 Computer simulation of natural selection

Evolutionary theory can be broken down into two major component theories. Firstly there is descent with modification, that is to say organisms change over a period of time (a **kinematic theory**). Secondly, the process of change is due to natural selection (a **dynamic** theory). The kinematic theory can be derived in part from the study of the fossil record. The dynamic theory of natural selection was Darwin's major contribution to evolutionary theory. The mechanism of evolution, how natural selection works, is very important and is dealt with in the computer simulation entitled *Natural selection game.* You will need to ask your tutor for this.

5.4 Speciation

The variety of life on Earth has arisen as a result of the process of speciation, the derivation of two or more species from a previously existing species over a period of time.

Figure 41 illustrates **divergent speciation,** or divergence. This is the situation in which an ancestral species (species **A**) gives rise to two separate descendant species (species **B** and **C**). To understand the possible mechanisms which explain how divergence works, it may be helpful to look again at the species concept (see unit 1, *Inquiry and*

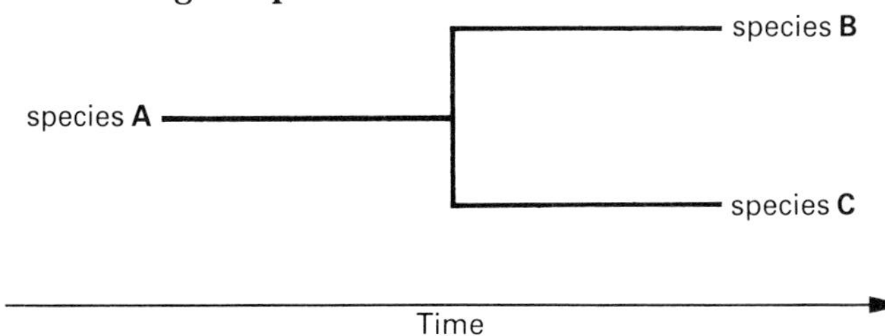

41 Divergent speciation

investigation in biology, section 4.6). Members of a species have the potential to interbreed and produce viable offspring. Thus, because they exchange genes, they belong to the same gene pool.

Members of closely related, but separate species, which live in the same region do not normally interbreed. If they do, the offspring are often infertile, or have very low fertility (for example horses and donkeys may mate and produce infertile mules or hinnies). The gene pools remain separate, and there is no gene flow between them. This integrity of species forms the biological basis of speciation, and provides a clue to the mechanism of evolution.

SAQ 64 Look at figure 42. How do you account for the existance, in Europe, of the two similar, but distinct, species of gull, the herring gull and the lesser black-backed gull.

42 A ring species of gulls surrounds the North Pole. The change between neighbours is gradual, but in Europe the two ends of the ring are distinct species

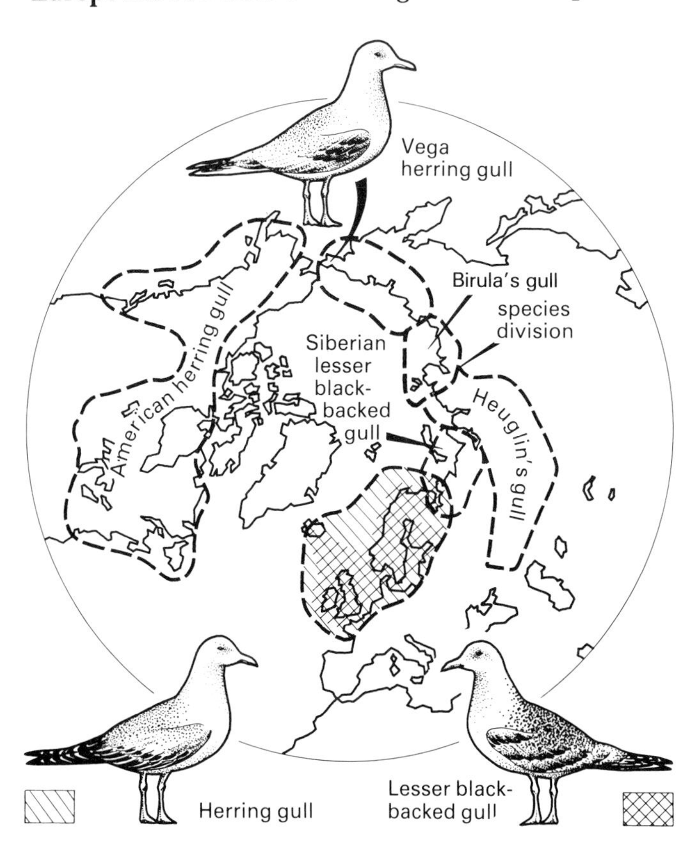

We shall now go on to consider the various ways in which divergence occurs, that is the mechanisms of divergence.

5.4.1 Allopatric speciation

This mechanism involves some type of natural barrier coming between parts of the ancestral population.

The first step towards divergence is taken when groups of organisms within a population become separated from each other so that the gene flow is either restricted or stopped. The two groups are prevented from interbreeding because of the physical barrier between them. For example, this could be a stretch of water or a mountain range. Over a long period of time, the two populations would eventually produce different phenotypes. This is because of the recombination of genes and the action of natural selection. Each population becomes adapted to its local conditions. In fact, the morphological differences may eventually become so great that they are no longer able to interbreed.

This process is known as geographical isolation, and it is the main step towards allopatric speciation.

5.4.2 Parapatric speciation

This occurs when a genetically unique group of organisms arises in a population and occupies a niche within the same environment as the rest of the species. The divergence from the main population occurs because this new group is reproductively isolated from the rest. It is thought that random chromosome changes occuring in some organisms produce rapid physiological and reproductive differences from the main population.

5.4.3 Polyploidy

This is sometimes known as abrupt speciation because of the speed with which it occurs. Polyploidy is said to have occurred when a cell increases its chromosome complement, usually by undergoing mitosis without cytokinesis (see unit 2, *Cells and the origin of life*, section 8.3). Polyploid organisms cannot usually reproduce with the ancestral diploid forms because the uneven chromosome number arrests cell division and subsequent development of the embryonic cells. However, polyploid organisms can often self-fertilise, or fertilise with other polyploids with the same chromosome number. Reproductive isolation can, therefore, occur in a single generation. A new species can be formed if the polyploids can produce viable offspring and are able to find a niche in which to live. An example of this is *Spartina anglica* which was discussed in section 5.3 of the unit *Genetics*. Polyploidy is far more common in plants than in animals. It has been estimated that over 40% of angiosperm species have arisen by polyploidy.

SAQ 65 For divergence to occur, the species must be reproductively isolated. Explain the meaning of the term reproductive isolation.

5.5 Reproductive isolation mechanisms

The following sections cover the ways in which interbreeding is prevented.

5.5.1 Temporal isolation

The organisms breed at different times of the year (figure 43). This is very common in plants, but also occurs in some insect groups, molluscs and amphibia.

43 Temporal isolation

5.5.2 Ecological isolation

In this case organisms live in different habitats (figure 44). An example of this can be seen in the mouse deer, *Peromyscus* sp. This species has one population living in a woodland environment and another occupying the plains. Under natural conditions the two populations remain within their own habitats, and so never come together to reproduce. They *will* breed if brought together, however.

44 Ecological isolation

5.5.3 Behavioural isolation

Here, the organisms have different behaviour patterns (figure 45). Some patterns of behaviour are known to be inherited. Those concerned with courtship and mating are very important in this context. In *Drosophila*, for example, normal mating behaviour involves a definite sequence of wing and body actions. An experiment showed that two very closely related species of *Drosophila* would never normally mate because the courtship 'dance' of the

45 Behavioural isolation

male was different. However, if the antennae of the female were removed, she would permit mating. It was concluded that she was unable to detect 'wrong' courtship patterns and so allowed mating.

SAQ 66 What criticisms can you make of this assumption?

5.5.4 Gametic isolation

Any slight change in genetic code will first be manifested by change in proteins. This can have immediate and profound effects on the physiology and biochemistry of the cells, including the gametes. Even if mating occurs, fertilisation may be blocked because of the incompatability of the gametes (figure 46).

46 Gametic isolation

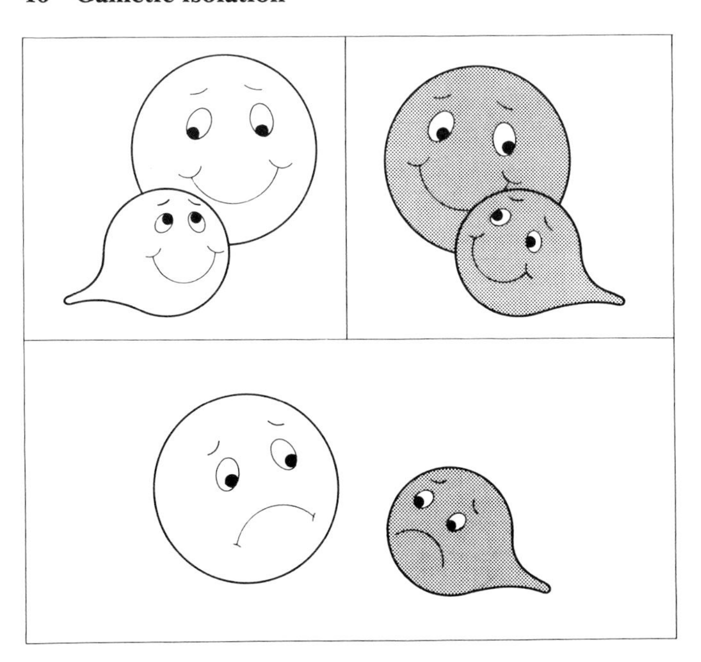

5.5.5 Mechanical isolation

This occurs when closely related species are unable to mate because of anatomical differences which make it physically impossible to bring the gametes together (figure 47). Domestic breeds give an example of this type of isolation. It would be unlikely that a Shire mare and a Shetland pony stallion could successfully mate, even though they are of the same species.

47 Mechanical isolation

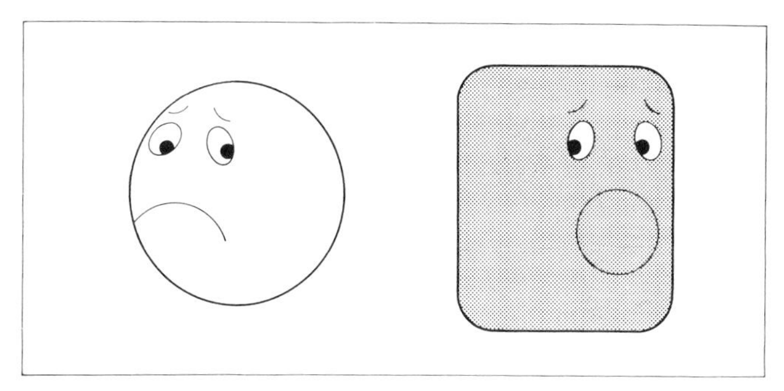

SAQ 67 What would be the likely outcome if the only dogs left in the world were either St Bernard's or chihuahuas?

5.5.6 Hybrid isolation

Hybrids often die soon after fertilisation, or at some subsequent stage of embryological development (figure 48). Even if they reach maturity, hybrids are often sterile. This is due to differences in the chromosome number of the parents. It means that meiosis cannot occur properly in the hybrid's gonads, and so viable gametes are not produced.

48 Hybrid isolation

If a hybrid *is* able to produce viable offspring, its overall fitness is often reduced, and the number of organisms in the F_2 generation may be less than the number in the hybrid F_1, and so on.

5.6 Extension: Galapagos finches

During his voyages on the Beagle, Darwin visited the Galapagos Islands which are about 1000 km off the northwest coast of South America. Darwin realised that the birds on the Galapagos Islands were different from those of the mainland. He counted 13 species of finch, and noticed that each had a slightly different shaped bill.

SAQ 68 In figure 49, which bill is adapted to which food?

Natural selection only partly explains why this adaptive radiation took place in the finches of the Galapagos. The geographical isolation mechanisms operating in the islands also played an important part in the evolution of the finch species.

SAQ 69 What main factors have contributed to the evolution of so many finch species in the Galapagos?

SAQ 70 The following statements relate to the diagrams in figure 50. Write out the statements in the correct order and, beside each, put the letter of the diagram which corresponds to it.

1 The finches increased in numbers and, under the influence of natural selection, gradually became adapted to the local environment.

49 Galapagos finches and their foods

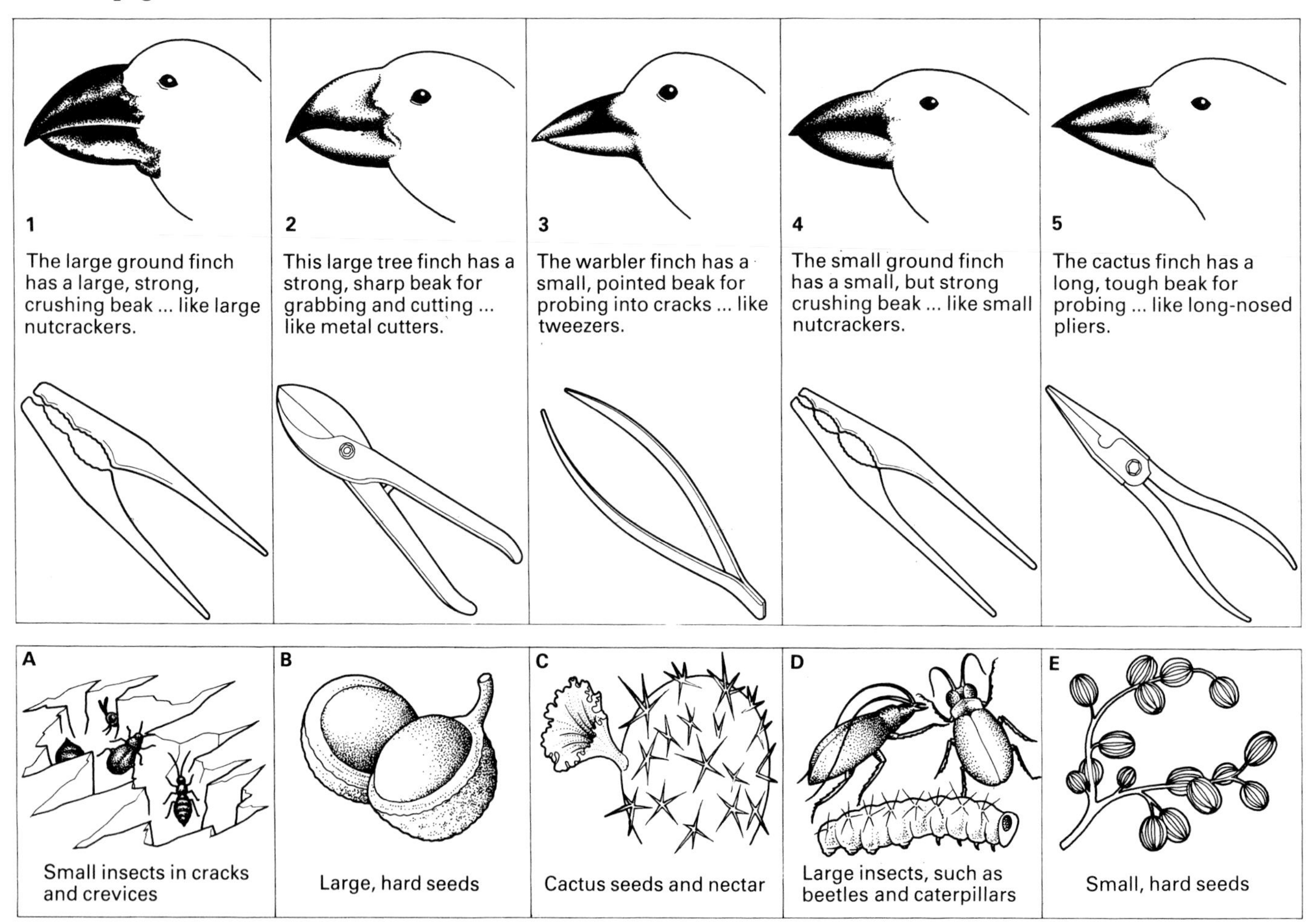

2 Some finches from the second island managed to fly back to the first island, but reproductive isolation had occurred between them and the existing population.
3 Originally, there were no finches on the islands. Some finches from the mainland managed to fly across to them.
4 Some of the finches managed to fly to a second island where the environment was different.
5 This process was repeated over and over again as the finches colonised more and more of the islands.
6 Adaptation to the conditions on the second island gradually took place.

5.7 References and further reading

Neo-Darwinism by R. J. Berry, Institute of Biology Studies No. 144.
The Evolution of Melanism by H. B. D. Kettlewell.
The Theory of Evolution by John Maynard Smith (3rd edition).
Origin of Species BM(NH) 1981.
Darwin's Missing Evidence by H. B. D. Kettlewell.
Evolutionary Principles by Peter Calow.
Splendid Isolation by George Gaylord Simpson.

50 Colonisation of the Galapagos by finches

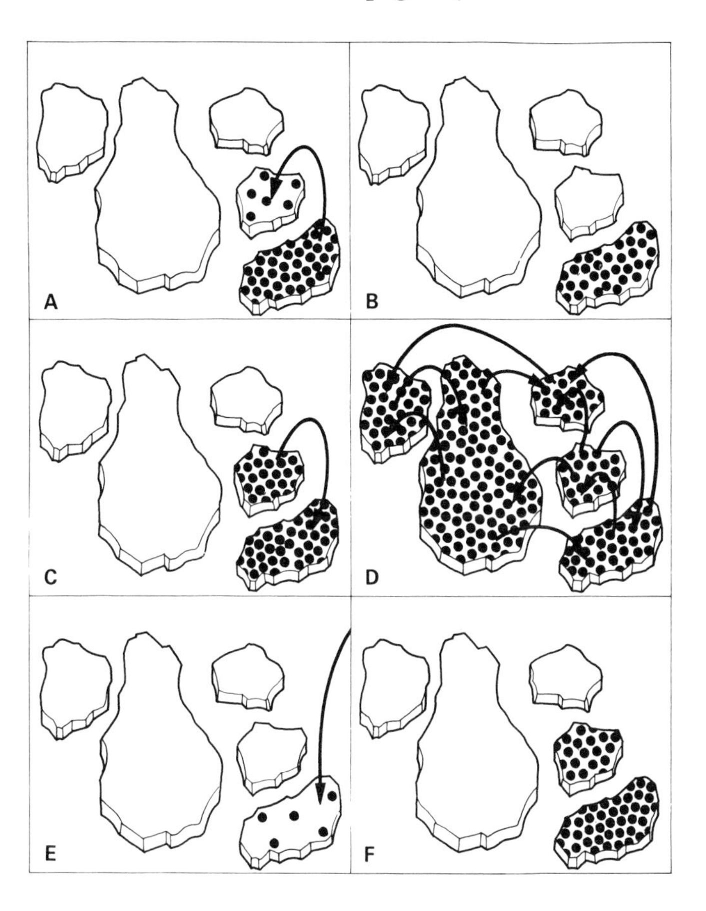

Section 6 Interactions

6.1 Introduction and objectives

The purpose of this chapter is to indicate something of the range of types of interactions between organisms and to map out the broad categories. Some of the examples will already be familiar to you from work in previous units, others may be completely new. In either case, space will not permit detailed descriptions, and so you are encouraged to follow up some of the references given at the end of the chapter.

At the end of this section you should be able to do the following.

(*a*) Distinguish between interactions within populations and interactions between populations.

(*b*) Explain the advantages of a multicellular construction.

(*c*) Distinguish between individuals and colonies, and list some of the advantages and disadvantages of each form.

(*d*) Give examples of interactions between members of the same, and of different, sexes.

(*e*) Distinguish between populations and societies.

(*f*) List some of the features of a social group.

(*g*) List the types of interaction occurring between populations.

(*h*) Define and give examples of competition, neutralism, mutualism, predation, commensalism, amensalism and different forms of mimicry.

(*i*) Explain why allocating organisms to such categories is sometimes problematic.

(*j*) List, with examples, the types of semiochemicals.

(*k*) Describe the life-cycle of a parasite.

(*l*) Outline the possible evolutionary origin of eukaryotic cells.

(*m*) Give examples of coevolution.

6.2 Types of interaction

By now you will be aware that organisms cannot live in isolation; all organisms depend to a greater or lesser extent on a range of other organisms. From our knowledge of ecosystems and nutrient cycling (such as the carbon or nitrogen cycles) one could argue that all organisms on this planet are interdependent. However, we shall be concentrating on the relationships of organisms that are found together, that is organisms that are **associated**. Three main criteria will direct the way we create our categories:

(*a*) whether the interactions are beneficial or adverse;

(*b*) the extent to which the association is a necessary or **obligate** one;

(*c*) whether the organisms are of the same species.

For the sake of convenience, we shall consider interrelationships under three main headings.

Interactions within populations will consider intraspecific relationships. *Interactions between populations* will consider interspecific relationships. The third section, *Interactions between taxa*, will adopt a more historical perspective and consider certain long-term evolutionary associations.

6.3 Interactions within populations

The evolution from unicellular to multicellular construction has had certain demonstrable advantages. These include:

(*a*) the ability to replace damaged or worn out cells and hence to live longer,

(*b*) differentiation of cells leading to greater functional efficiency and reproductive success,

(*c*) increase in size tending to physiological stability, and

(*d*) the construction of different forms of architecture.

The enormous diversity of form that we encounter in multicellular organisms leads to certain interesting questions, as we shall see.

6.3.1 The concepts of the individual and the colony

As humans we are individuals, and we also think of many of the common animals and plants around us as being individual creatures, such as dogs, cats, ladybirds, snails, oak trees and so on. Equally, just as we can talk about interaction between human beings, we can observe, study and hypothesise about interactions between individual cats, dogs, ladybirds and so on. However, as you will have discovered, the concept of an individual is often very difficult to apply to such organisms as grass, bracken, liverworts, slime moulds, fungi, filamentous algae, sponges and coelenterates. Why should this be so? Partly this may be because we bring to biology certain assumptions.

One of these is that animals are discrete individuals, physiological units with a central organising principle (a brain). However, when we come to study the diversity of life, we discover that many of the creatures that we want to call animals have no obvious central organisation; indeed, for that very reason, many animals like coral and sponges were initially classified as plants.

By now, you will have realised that plants, in general, show far more plasticity of form than do animals. For instance, the number of leaves on a clover plant is usually three, but four is not a catastrophe. Equally, we can cut the odd branch off a tree without causing any great harm. On the other hand, comparatively minor changes to an animal can prove disastrous. (This plasticity of form in plants is, in part, attributable to their pattern of growth which continues throughout life and is responsive to environmental factors in ways that animals are not.)

Plants have evolved a considerable array of means for asexual propagation: stolons, runners, rhizomes, tubers, corms, bulbs, fragmentation, gemmae, spores, soredia, isidia and so on. Many of these means of reproduction, like rhizomes and stolons, make it an arbitrary matter to distinguish where one plant ends and the next begins, particularly as all the cells will be genetically uniform. Therefore, the concept of the individual is frequently both difficult to apply to plants in practice and of limited value. Accordingly, many ecological studies on vegetation often focus on features like frequency and percentage cover rather than numbers of individuals per unit area (density).

Similar problems arise in a number of lower animal phyla. Animals like sponges, show cellular differentiation. There is, therefore, collaboration between cells (for example feeding currents are set up) but no central coordination. Like plants, portions of the sponge may be removed without harm but, unlike plants, if the sponge is completely reduced to component cells (by chemical treatment and sieving) the cells can reaggregate to form a functioning sponge. Furthermore, mixtures of cells of different species reaggregate correctly with only cells of the same species coming together.

In coelenterate groups individuals may occur as polyps like *Hydra* (see unit 2, *Cells and the origin of life*, sections 8.6 and 8.7) or as medusae, that is jellyfish.

Many polyp-forming coelenterates form sessile colonies in which the polyps may be differentiated for feeding or reproduction and sometimes protection. The individual polyps act as an association of sub-individuals though they are not discrete physiological units. In the case of the Portuguese man-of-war, *Physalia* (figure 51), tentacles up to 20 m long are composed of masses of

51 ***Physalia***

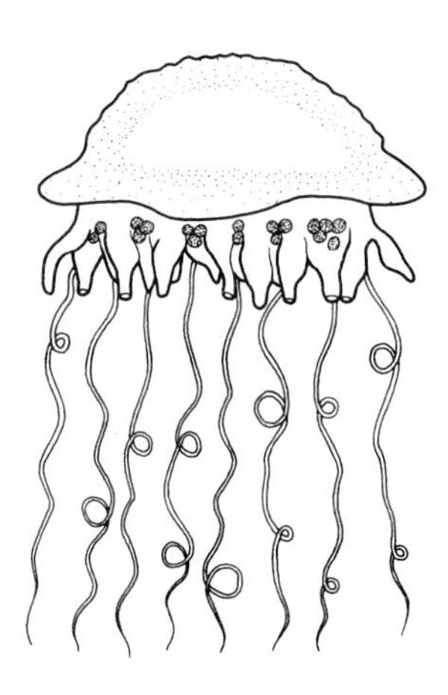

polyps *and* medusae hanging from a gas-filled float. In *Physalia*, the collaboration of sub-individuals is so complete and so necessary that it is not obvious whether we should call the 'man-of-war' a colony or an individual. (This is just one example of a blurring of categories. We shall meet more later when we re-examine some other conventional terms like parasite and symbiosis.)

6.3.2 Intraspecific interactions between individuals

Most colonial animals are sessile; *Physalia* is an exception, though it too can be seen as a sessile colony hanging from a bag. In contrast, most individual animals locomote in order to seek food, reproduce, escape harmful conditions, and so on. This has led, by and large, to a distinct anterior–posterior construction, cephalisation and the development of a central nervous coordinating system; in other words, a distinct individual organism. Being an individual rather than part of a colony has disadvantages as well as advantages. Being an individual can mean isolation, principally reproductive isolation. Thus, the one obligatory intraspecific interaction for most individuals of dioecious species will be between males and females.

6.3.3 Male–female interactions

For many species mechanisms are necessary for seeking and identifying a mate, and the more dispersed members of the population are, the more sophisticated the mechanism will need to be. Some organisms have solved this problem in unusual ways. The angler fish, *Photocorynus spiniceps*, roams the ocean at great depths and over a huge area. Since finding a mate at such depths would be virtually impossible, each female, when young, picks up a male that remains both small and permanently attached to the female's head. The nematode worm *Trichosomoides crassicauda* parasitises the bladder of rodents. This too is an unprepossessing environment in which to seek a mate. Accordingly, each female *Trichosomoides* carries a male in her uterus, where it remains.

Such examples are exceptions, however, and most species have had to confront the problem in another manner. A key problem is to avoid undue gamete wastage. Shedding gametes into water is wasteful, but can be successfully employed by species whose members are not too widely dispersed. In such organisms, a genuine interaction may be almost non-existent. However, most species have evolved some form of mate-seeking/selection mechanism. This may be an indirect process brought about by environmental factors or a true seeking/identification mechanism. Consider, for instance, the contrasting examples of the palolo worm and the silk moth. The palolo worm, a burrowing annelid, sheds its gametes into the sea. The chances of fertilisation being successful are increased because release of the gametes is synchronised. These worms depend on the phases of the moon for coordinating their sexual activities. Spawning in enormous numbers occurs at dawn on the day before the moon enters its third quarter in October and November. Each worm responds individually to this external environmental signal and so all spawn at the same time. These worms are a delicacy and fishermen know they can only be caught at these particular times of the year.

In the case of the silk moth, the female secretes minute quantities of the alcohol bombykol from glands on her abdomen. The male moth has large feathery antennae which bear sensory hairs that respond to bombykol (figure 52). Even at very low concentrations (14000 molecules per cubic

52 Male silk moth antenna

centimetre of air) the male can respond to bombykol and will be attracted to the female even from distances of over a kilometre.

Bombykol is an example of a **pheromone,** a very important group of chemicals that we shall consider later (see sections 6.3.5, 6.4.3 and 6.4.5).

Mating is also commonly preceded by courtship which may be lengthy and complex, as in certain social creatures, or quite perfunctory. One very important function of courtship is to ensure the prospective partner is of the correct species. For instance, many species of *Drosophila* may be quite similar but different species have distinct signals. Sometimes, a complex courtship dance may precede mating, thus enabling the female to check that the male knows the correct dance and is, therefore, a member of her species. In a group of Hawaiian 'picture-winged' *Drosophila*, male flies display rather like peacocks. The wing-markings elicit reproductive behaviour only from females of the correct species.

Apart from olfactory and visual signals, tactile and acoustic signals may also form part of the communication between potential partners. In woodlice, a brief courtship goes as follows. The male seems to detect a receptive female by scent. He stops and, moving his antennae about, rests them on the female. If not rejected, he climbs on her back, licking her head with his mouthparts and drumming on her back with his front legs. After this, mating ensues.

Much better known is the singing of crickets. Male crickets sing when they are ready for copulation: this behaviour seems to be induced by a chemical in the sperm. The purpose of the sound is to attract a mate and when a female comes within sight, the calling song changes to a courtship song. As the female approaches, antenna contact is made. Then the female turns around, the male climbs on her back, stops singing and they copulate.

6.3.4 Male–male interactions

Up until now, we have only considered female–male interactions, but the crickets also have distinct male–male interactions. The calling song referred to previously has a secondary effect. Males will tend to move away from a nearby male until the rival's sound is sufficiently diminished. (If the sound intensity is 100 dB at the source, the rival may move until the level is about 40–50 dB.) This has the important effect of spacing out the males and establishing territoriality. In the event of a rival getting too close, a third type of song is given accompanied by an aggressive visual display.

Many animals lay claim to territories and mark it by various behaviours, scents, and the like, defending it against encroachment by, usually, other males. Ring-tailed lemurs have special glands for scent-marking and 'stink fights'.

Some male–male interactions may not be quite so obvious as aggressive behaviour but, nevertheless, just as important. The members of the genus *Blops* are large black beetles which live in dry habitats in certain Mediterranean countries. The males of some of these beetles have a gland on the ventral surface of their abdomen that produces a jelly-like material. This substance is transferred to the female during mating. This seems to mark the female chemically to deter subsequent males, a sort of anti-aphrodisiac pheromone.

No doubt you will be able to think of many other examples of male–male and male–female interactions. What female–female interactions can you list?

6.3.5 Other intraspecific interactions and pheromones

Up until now, we have only considered intraspecific relationships and interactions based directly or indirectly on sex. Yet members of a species may influence other members for a variety of reasons and in a number of ways. We shall consider a few examples mediated by pheromones. **Pheromones** are chemical substances produced by an organism which influence other members of its species. Chemicals may be produced which affect other species, but these are called allelochemics and will be discussed later (see sections 6.4.3 and 6.4.5).

Pheromones are not exclusive to animals. Brown algae may produce sex pheromones, for instance *Fucus serratus*, the serrate wrack of rocky shores, produces fucoserraten. This substance attracts the motile gametes to the female conceptacles.

Pheromones are not restricted to terrestrial animals, aquatic animals may also produce them. The moulting hormone, crustecdysone, of some crabs seems to double as a sex pheromone. The sea slug *Navanax* secretes a yellow oil when disturbed. This oil induces other *Navanax* sea slugs to move away. The sea anemone *Anthopleura elegantissima* produces anthopleurine when disturbed, which causes other *Anthopleura* nearby to contract. In other words, anthopleurine is an alarm chemical fulfilling the same sort of function as the blackbird's 'pink-pink' warning call.

Pheromones are also used by termites to lay trails, but this brings us to a new level of complexity of interaction and to the concept of the society.

6.3.6 Populations and societies

From unit 9, *Ecology*, section 3, you will recall that a **population** is defined as a group of organisms of the same species that occupy a particular area. Since such members will be capable of freely interbreeding, the population will normally have genetic continuity.

We have seen that activities of members of a population may appear coordinated due to some environmental effect, such as flowering seasons of plants and palolo worm breeding periods. We also saw that individual organisms can influence other individuals in a number of ways, either chemically or by some other means of communication. Communication (defined as an action by an organism that may affect the behaviour of another in an adaptive manner) leads to the possibility of collaborative behaviour and to the development of social hierarchies, in short, to the formation of societies. A population may be a society but need not be. Societies are delineated by patterns of communication. A population may be comprised of several societies, but since social grouping tends to restrict breeding, a population divided into societies may rapidly become several populations, that is gene flow between groups may become reduced.

Cooperative behaviour in social organisms can lead to differentiation; this may be simply a question of roles. Male and female birds may have particular roles in defence of territory, nest-building, feeding and protection of young. In other groups, particularly the social insects, there may be differentiated castes, such as worker bees or soldier termites. In these cases, the proportion of each class in the population can affect the fitness (reproductive success) of the group. In turn, this may affect the number of individuals of different ages. That is why social groups tend to have stable or, at least predictable, age distributions, that is **demographies.** Differentiation of roles, evolution of castes and distinct demographies are not the only features of societies. Members of societies also show varying degrees of cohesiveness (how much they tend to remain close together for protection, feeding and communication and so on). This grouping can also vary in the ways social groups are connected. For instance, schools of fish and flocks of birds may have quite simple organisations, but groups like monkeys may have very complex hierarchies. Having a hierarchy may increase the degree of coordination and the efficiency of the group.

The more socialised the groups are, the more time members tend to spend devoted to the groups. For instance, lemurs have a simple social organisation and seem to devote about 20% of their time to social acts. While the macaques, which have a more sophisticated social structure, may spend as much as 80–90% of their time in social acts.

Social evolution is itself a fascinating topic. How did societies evolve? Is social evolution reversible? Can social species become solitary? Finally, a key issue is that of altruism and its relationship to group selection. This has been referred to briefly in section 1.3. A new field of biology that deals with these topics has begun to emerge, called **sociobiology,** and the interested student is recommended to a seminal book in this field, *Sociobiology* by E.O. Wilson. However, as a new research field, there are inevitably a number of current controversies which, of course, makes the topic exciting. Perhaps it is important to repeat an earlier warning. Altruistic behaviour in animals is not necessarily the same thing as altruism in human beings and that biologists who claim to provide an evolutionary or scientific basis for human ethics may be making claims outside their real field of expertise.

6.3.7 Concluding remarks

We have looked briefly at aspects of the relationship existing between sub-individuals of a colony, individuals, and members of societies. We have considered some of the main forms of communication, chemical, visual, audible, and so on, though we have studied nothing of the nature of that communication, particularly audible communication in social organisms and the emergence of language. We have seen that the nature and purpose of interrelationships is varied and may be complex. This chapter should, therefore, be considered as little more than brushing the surface of an enormous field of literature under a variety of headings: sociobiology, animal behaviour, ethology, animal language, and so on.

6.4 Interactions between populations

In this section, we shall consider interactions between populations. You will already be familiar with many examples of interaction between members of a community (from *Ecology*, unit 9). An interaction between two populations may be beneficial, adverse to either side or neither. If we use symbol + to denote beneficial and − to denote adverse, we can list the various possibilities in a table (see figure 53). We shall review each of these categories in turn. It should be noted that two of them are extremely important, competition and predation, but since these will have been touched on in several previous units such as unit 3, *Energy and life*, on feeding and nutrition and unit 9, on ecology, very little attention will be paid to these topics here.

6.4.1 Competition

Competition for resources may be intraspecific (between members of the same species) or interspecific (between individuals of different species). In either case, the ratio of demand to supply when competition is keen is believed to be a very potent evolutionary force, though it is notoriously difficult to demonstrate in natural communities and usually indirect evidence is used to infer competition.

We can distinguish two broad types of competition.

53 Summary of the various sorts of interactions that may occur between two populations

	Species		
Type of interaction	*A*	*B*	*Nature of the interaction*
Competition	—	—	Each population inhibits the other
Neutralism	0	0	Neither population affects the other
Mutualism	+	+	Interaction is favourable to both but is obligatory
Protocooperation, Mullerian mimicry	+	+	Interaction is favourable to both but is not obligatory
Predation	+	−	Population A, the predator, kills and consumes members of population B, the prey
Parasitism, Batesian mimicry	+	−	Population A, the parasite, exploits members of population B, the host, which is affected adversely
Commensalism	+	0	Population A, the commensal, benefits whereas B, the host, is not affected
Amensalism	−	0	Population A is inhibited, but B is unaffected

Interference competition is used to denote direct interaction, such as aggressive encounters between animals or allelopathy between competing plant species. (**Allelopathy** is the secretion of chemicals by plants that prevent others growing around them.) Interference competition may have a real effect on the population even when the resource is not, in fact, limited. The term **exploitation competition** covers situations where the competitors do not actually meet. An example of this is competition between moles and blackbirds for earthworms.

6.4.2 Neutralism

Strictly, neutralism would be the existence of two populations in a community that do not interact. In practice, neutralism probably never arises because there are bound to be indirect interactions between all the populations of a community.

6.4.3 Mutualism

When both populations benefit from an association that is obligatory, we call this mutualism.

Before we continue, a brief word about some of these terms might be useful. When we try to categorise objects or events in the natural world, we frequently find that, sooner or later, we discover something that does not fit our scheme, perhaps because we have tried to impose a set of distinct categories onto something which is much more of a continuous spectrum. Types of interaction is a case in point and, over the years, different authors have used terms in different ways in an attempt to improve matters. For instance, some books may use the term **symbiosis** (meaning living together) to cover all interspecific interactions except predation. Others have reserved the term for all such interactions except parasitism, and yet others have restricted the term to cover only quite intimate associations. The scheme we use here is quite modern and comprehensive and you should have no difficulty adapting to it.

To return to mutualism (or symbiosis in a restricted sense), why is it obligatory? The answer is that members are usually physiologically interdependent. A frequently cited example of mutualism is the lichen (figure 54). Lichens are composed of a fungus and an alga. Clearly the relationship is more than the sum of the parts. Lichens show (*a*) structural modifications (such as features of the thallus and the reproductive structures called soredia), and (*b*) physiological activities (such as the formation of so-called lichen acids) that are found in neither component when isolated. Indeed, lichens have, in effect, their own names distinct from their components (though the name is strictly that of the lichenised fungus). However, the true nature of the relationship between the fungus and alga has long been a vexed question, and it is certainly possible that for some lichens it is a matter of an alga being parasitised, while in other partnerships the relationship is more mutualistic. Indeed, in some cases, it looks as if it is the alga that is the dominant partner.

54 Section through a lichen thallus

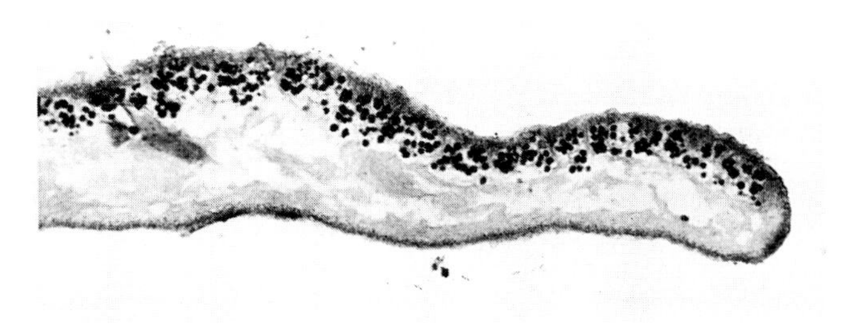

Other examples of mutualism are widespread. Many aquatic invertebrates possess symbiotic algae. Such animals include green hydra and the reef-forming corals among the coelenterates. Also *Paramecium bursaria*, certain molluscs and sponges contain algal endosymbionts. In analysing the relationship, the possible advantages to the heterotrophs are usually clear: a flow of carbohydrate, an oxygen source, possibly conversion of nitrogenous waste, vitamin synthesis, and so on. The advantage to the autotroph (as with the lichen) is less clear-cut. Protection is usually suggested, and in the case of lichens it is true that the alga can live in the thallus in nutrient-poor conditions. In the case of invertebrate hosts, perhaps the relationship looks a little more like internal farming, especially when one considers the unusual phenomenon of **chloroplast symbiosis**. Certain marine molluscs feed on seaweed but incorporate the chloroplasts into the cells lining their gut. Here the chloroplasts may live and photosynthesise for some days while releasing carbohydrates to the mollusc.

Apart from heterotroph–autotroph relationships, we can find examples of autotroph–autotroph and heterotroph–heterotroph interactions. Here are some examples.

Anabaena is a member of the blue-green algae, a primitive group of plants that are similar in some respects to bacteria. *Anabaena* is found as an endosymbiont in the floating fern *Azolla*. The endosymbiont *Anabaena* sacrifices its ability to photosynthesise but acts as a very efficient nitrogen-fixer, allowing *Azolla* to live in very nutrient-poor waters. *Nostoc*, another blue-green alga is found in swellings in the stem of *Gunnera*, a flowering plant that looks like monster rhubarb. Like *Anabaena*, the *Nostoc* also acts as a nitrogen fixer.

Many mutualistic relationships between animals can be found, but the following example is fascinating for a number of reasons. Firstly, it is an example of a relationship between several organisms not just two. Secondly, it is highly suggestive of a view that eukaryote organisms have evolved through symbiotic relationships between prokaryotes. Let us see why.

Mixotricha paradoxa (figure 55) is a flagellate that lives in the gut of the Australian termite *Mastotermes darwinensis*. *Mixotricha* helps the termite digest wood and cannot exist outside the termite gut. This is, of course, an example of mutualism, but it is *Mixotricha* itself that is so fascinating. Each *Mixotricha* cell is a collaboration of five types of organism. *Mixotricha* is a flagellate, but its flagella do not contribute to its movement. Rather the animal is propelled by large and small spirochaetes that are attached to the

55 *Mixotricha paradoxa* has three cortical symbionts (two types of spirochaetes and one bacterium) and one type of endosymbiotic bacterium

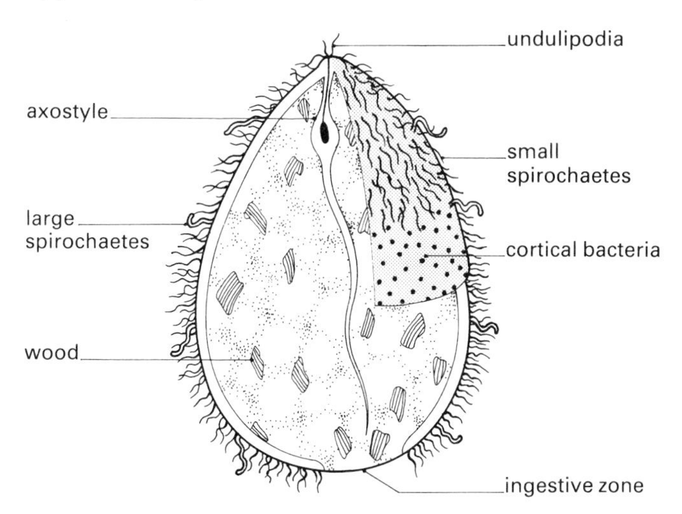

surface. At the point of attachment of the spirochaetes are bacteria which may have some sort of anchoring function. Finally, another sort of bacterium is also found inside *Mixotricha*, and it is believed these may serve the function of mitochondria.

Since symbiosis is very common, Margulis has suggested a sequence of stages from irregular associations at the host surface through motility symbiosis (like the spirochaetes on *Mixotricha*) to 'emboitement' or **incorporation.** Margulis has proposed that a process like this enabled the evolution of eukaryotic cells, and that flagella, mitochondria, plastids and even the nucleus owe their origin to symbiosis.

Examples of mutualism can be found even including viruses. **Paramecium aurelia** has a λ-virus without which it cannot synthesise folic acid.

Finally, let us look briefly at the role of micro-organisms in chemical interaction. This is an interesting area and might make us question whether we should think of mammals as partners in a symbiotic relationship.

It has been known for a long time that micro-organisms are partly responsible for odours like bad breath and axillary (under-arm) odour. More recently, it has become clear that some odours may have a biological significance, that is to say the odours may be acting as pheromones. We shall consider just two examples.
Rhesus monkey females, *Macaca mulatta*, produce volatile fatty acids in their vaginal secretions. These fatty acids are produced by microbes but cause nearby males to become sexually active; the chemicals have therefore been termed **copulins**.

The red fox, *Vulpes vulpes,* has a special fermenting system. Two anal sacs (each opening to the skin surface on either side of the anus) contain a microflora which determines the characteristic odour raised by the fox.

In these latter two cases, we may imagine the mammal surviving without assistance of this microbial flora, but whether they would survive and breed satisfactorily we have no way of knowing at the moment. Therefore, we cannot be sure if such cases are really mutualistic or rather proto-cooperative, but whatever we decide the line between the two categories will be very fine.

6.4.4 Protocooperation and commensalism

Protocooperation covers relationships that are non-obligatory and looser than mutualistic ones. Pollination of plants by animals and dispersal of their seeds include examples of protocooperation. The benefit to the plants is obvious and the animals may feed on nectar or a fleshy fruit, and so on. It must be said that in some cases the modification of the plants is so extreme that only a single species may be attracted or may fulfil the role as pollinator. In these circumstances, the relationship has crossed the boundary to mutualism. For example, *Yucca* can only be pollinated by the moth *Pronuba yuccasella*, and where the moth does not occur *Yucca* never sets seed.

Just as the distinction between mutualism and protocooperation may be a fine one, it is not always clear where protocooperation ends and commensalism begins. Strictly, protocooperation benefits both members, but the advantage to one may be very slight indeed. The remainder of this section refers to **commensalism**, in which one species benefits and the other is unaffected, but several of the examples could almost equally well be classed as protocooperation.

We shall consider commensalism between animals and we can identify four broad groups, firstly, commensalism based on cleaning. Oxpeckers, *Bucephalus erythrorhychus*, removes ticks and fleas from the skin of cattle, zebra, giraffes, and others, and the Egyptian plover, *Pluvianus aegyptus*, cleans the teeth of the Nile crocodile. Similar relationships have also been seen in fish, while, closer to home, the small, blind, white woodlouse *Platyarthrus hoffmanseggii* lives in ant nests (especially the British yellow ant, *Lasius flavus*) where it feeds on the ant excreta.

Secondly, commensalism may be based on protection and camouflage. For instance, the hermit crab, *Dardanus*, may carry on its shell anemones, *Calliactis* sp., which seem to protect it from octopuses.

Thirdly, commensalism may be synoecious (*syn* meaning together, *oikos* meaning house). This covers associations where an animal occupies the shell or burrow of an animal of another species. The mussel, *Mytilus edulis*, sometimes shares its shell with the pea-crab, *Pinnotheres pisum*.
Fourthly, commensalism may be based on transport where one species may use another to, in effect, hitch a ride! The remora, *Remora remora*, is well known. It clings to the underside of sharks using its specially modified dorsal fin. Close inspection of the legs of some invertebrates like harvestmen may reveal smaller invertebrates clinging on, such as mites.

Before going on to consider the remaining sections of the table shown in figure 53 there remains one further type of +/+ interaction (that is where both species benefit from the interaction), namely Mullerian mimicry.

During the course of evolution, prey species have evolved a number of strategies to reduce the likelihood of their being eaten. One of these methods is to become unpalatable either by making toxins or by feeding on plants and concentrating their toxins. Unpalatable or toxic animals then employ warning colouration to advertise their distastefulness: they use red, white, black or yellow. It is to the advantage of the prey (and to the predators) if unpalatable species share the same warning colours. Thus, unrelated species may bear a strong superficial resemblance to each other. This is known as **Mullerian mimicry**. Some non-toxic species may mimic toxic ones and thereby benefit; this is known as **Batesian mimicry.** Batesian (non-toxic) mimics are disadvantageous to the species they mimic (the distasteful model) because potential predators may encounter harmless mimics and thus take longer to learn to avoid the model. This, of course, brings us to the next section, the discussion of +/− interactions, that is where one species benefits at the expense of another.

6.4.5 Predation

The study of predation has become increasingly important in recent years as we have realised the need to know more about biological control of pest and disease organisms. Fortunately, predation has proved easier to observe and study than competition. Though this is not to say that solution to the problems, sought by applied ecologists and agronomists, have proved easier to find.

In predation the association is unequal: one population benefits, the other suffers. In competition, since both species suffer, avoidance mechanisms tend to evolve. In predation, avoidance mechanisms evolved by the prey are counteracted by more sophisticated hunting and trapping methods on the part of the predator. Prey species may develop alarm calls and camouflage and social species may have sentinels, but predators respond variously in ways including social hunting (as in lions and wolves), ambush (for example trapdoor spiders), rapid and accurate strikes (as in snakes, preying mantis), and so on.

Some authors employ the term predation to cover herbivory. There is, of course, an important difference between predation on animals and predation on plants and this relates back to the discussion about individuals in section 6.3.1. A predator kills an animal (except, maybe, colonial forms like coral) outright, but a herbivore will normally only consume part of a plant, which may recover. Herbivory is analogous to parasitism; compare, for example, a mosquito feeding on human blood and an aphid feeding on plant sap. (In section 6.4.6 we shall see that the distinction between predation and parasitism is not always easy to make). Since herbivory does not necessarily lead to the death of the 'prey', the selective pressures on plants will be different from those on animals. Nevertheless, plants have evolved an enormous array of antipredator devices. Crushed laurel leaves have been used by insect collectors for years to kill specimens. Laurel, some clovers, and bracken all produce cyanide if chewed or otherwise damaged. The cyanide (–CN) group, is attached to a sugar to form a glycoside. Also in the cells, but kept separate, is a β–glycosidase enzyme. If the cell is damaged these two substances come into contact and cyanide is released.

Plants have a variety of other antipredator devices too. Bracken, for instance, defends itself in the following ways. It produces cyanide like clover and other harmful chemicals which make it distasteful; these include phenolics, lignins, tannins, carcinogens and a sheep blindness factor. It also produces a chemical which mimics insect moulting hormones, an ecdysone. Bracken has periods when it is low in protein, so potential herbivores cannot depend on it throughout the year, furthermore, certain key amino acids are missing. Bracken produces a thiaminase enzyme. Last, but not least, bracken also has a protective association with ants. The ants feed on sugar produced at extrafloral nectaries (glandular hairs at intervals on the leaf) and, in return, they drive off many potential herbivores. Despite this armoury, over 20 species of phytophagous (plant-eating) insects have been found on bracken in this country. This illustrates the general point that selective forces on the prey are balanced by those on the predator.

However, there is a price to pay for these selective forces. In order to overcome obstacles presented by the prey, predators may become increasingly specialised as they become more adapted. In other words the range of possible prey may become restricted. Many insects are **monophagous**, that is depend on one food species; some are **oligophagous** and maybe feed on just one family of plants. **Polyphagous** insects can feed on a variety of plants. Insects that are polyphagous can often present a serious threat to crop plants. This is because they can feed and survive at low population levels on alternative, native plants and yet reproduce explosively when the crop monoculture becomes available.

The evolutionary consequences of predator–prey interactions are quite apparent in carnivorous associations too. We have discussed Mullerian and Batesian mimicry. Other mechanisms which have evolved include disruptive colouration (the shape of the animal is broken up by lines or blotches as in the tiger, leopard or rattlesnake), camouflage or cryptic colouration (for example, the *Biston betularia* example in section 5.3.1), flash colouration (some insects like grasshoppers have bright patches which attract predators' attention, when they stop moving the patch is concealed and the predator loses track) and alarm calls.

Before ending the section two further points must be included. Firstly, solutions to the ecological and agronomic problems mentioned at the start of section 6.4.5 usually require a quantitative analysis. Measurement of numbers of individuals, the changes over time, reproduction and mortality come under the heading of population dynamics. Since this is too large a topic to attempt here, references to further reading on this topic are included in section 6.6.

Secondly, this is a convenient point to clear up the terminology of chemical interaction since a number of examples have been given. The general term for a substance that carries information between organisms is a **semiochemical**. Substances acting between members of the same species are called **pheromones**. Substances acting between members of different species are either allomones or kairomones. **Allomones** benefit the emitting species, for example allelopathic chemicals and floral scents. **Kairomones** favour the receiving species, as in chemical clues which enable predators to recognise or locate their prey. Of course sometimes a substance may fall into several of these categories.

6.4.6 Parasitism

The imbalance in association in parasitism is the same as for predation: one species benefits, the other suffers. What distinguishes parasitism is that the prey does not die immediately. Parasitism is like a weak form of predation and analogous, as we have seen, to herbivory.

This section will make no attempt to describe life-cycles of particular parasites, though you may need to know something of these for your particular exam syllabus.

As we have seen before, associations between animals do not fit into convenient categories and different authors have had the greatest difficulty in defining what a parasite is. A simple definition is usually given as 'an animal or plant which lives partly or wholly at the expense of another living organism, its host'; but then so do predators! This definition has been expanded to 'that condition of life normal and necessary for an organism that lives on or in its host (which is usually a different and usually larger species) and that nourishes itself at the expense of the host without rapidly destroying it as a predator does its prey, but often inflicting some degree of injury affecting its welfare'.

Another definition relates parasitism to the degree of metabolic dependence of organisms on their host. In this context metabolic dependence includes nutritional requirements, developmental stimuli, digestive enzymes and control of maturation. Of course, if you consider each of these four in turn you will see an enormous difference between the circumstances and requirements of endoparasites and ectoparasites.

Yet another definition of parasitism emphasises the ecological relationship of the two populations. The features of this ecological relationship are:

(*a*) the parasite is physiologically dependent on the host,
(*b*) the infection process produces or tends to produce an over-dispersed distribution of parasites within the population,
(*c*) the parasite kills heavily infected hosts,
(*d*) the parasite has a higher reproductive potential than the host species.

The distinction between parasitism and mutualism can be blurred. Some mutualism can be seen as a special case of parasitism where the host benefits. We saw that lichens provide some ambiguous examples. Also the distinction between parasitism and predation is complicated by the existence of animals that have been called parasitoids; that is predators with parasite overtones! Parasitoids are usually Hymenoptera or Diptera which develop as larvae inside a host which is eventually killed. You will have met examples in unit 9, *Ecology*, when studying the holly leaf miner, *Phytomyza ilicis*, which is parasitised by any of nine species of wasp. Another example of a parasitoid is shown in figure 56.

Animal parasites come principally from the groups Protozoa, Platyhelminthes and Nematoda. Outside these groups we might cite leeches, ticks, fleas, mosquitoes and lice, but, depending on which definition we employ, these organisms could be classed as predators.

A key problem for all parasites is to establish the association in the first place. Commonly parasites have very complex life-cycles involving several hosts. *Schistosoma mansoni*, a platyhelminth that causes the disease bilharzia in humans, requires a snail as its alternate host (see figure 57). While another platyhelminth, *Schistocephalus solidus*, that parasitises freshwater birds in Britain, requires a crustacean and a fish as intermediate hosts (see figure 58).

The nature of the host–parasite interaction in

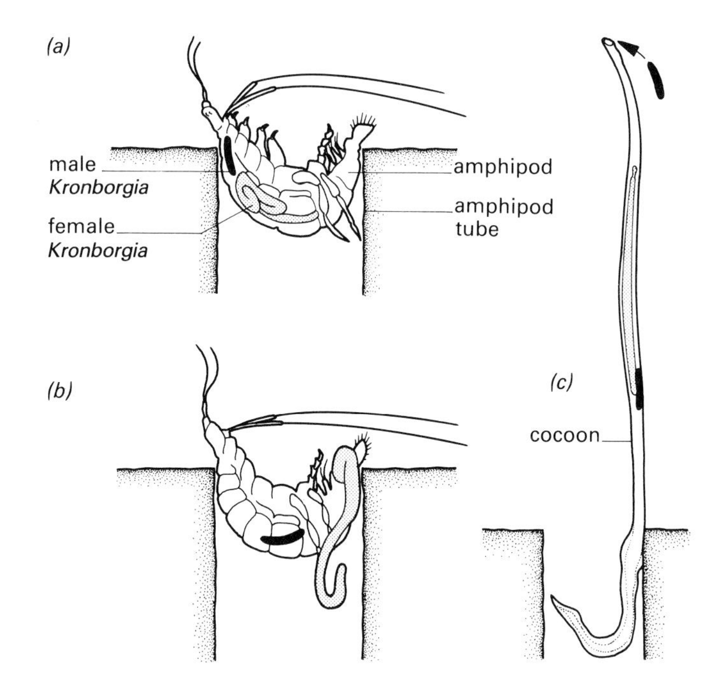

56 *Kronborgia amphipodicola*, a rhabdocoele platyhelminth parasitoid. (*a*) Gutless, ciliated males and females develop in the haemocoele of a tube-dwelling marine amphipod. (*b*) The worms release themselves from the crustacean, killing it in the process. (*c*) The female worm secretes a long tubular cocoon herself which is attached to the tube wall of the deceased host. Free-swimming males enter the cocoon and fertilise the female's eggs. She then spawns and dies leaving the cocoon filled with developing eggs. When the larvae hatch they swim to new hosts and penetrate their cuticles to complete the life cycle

57 The life-cycle of *Schistosoma mansoni*

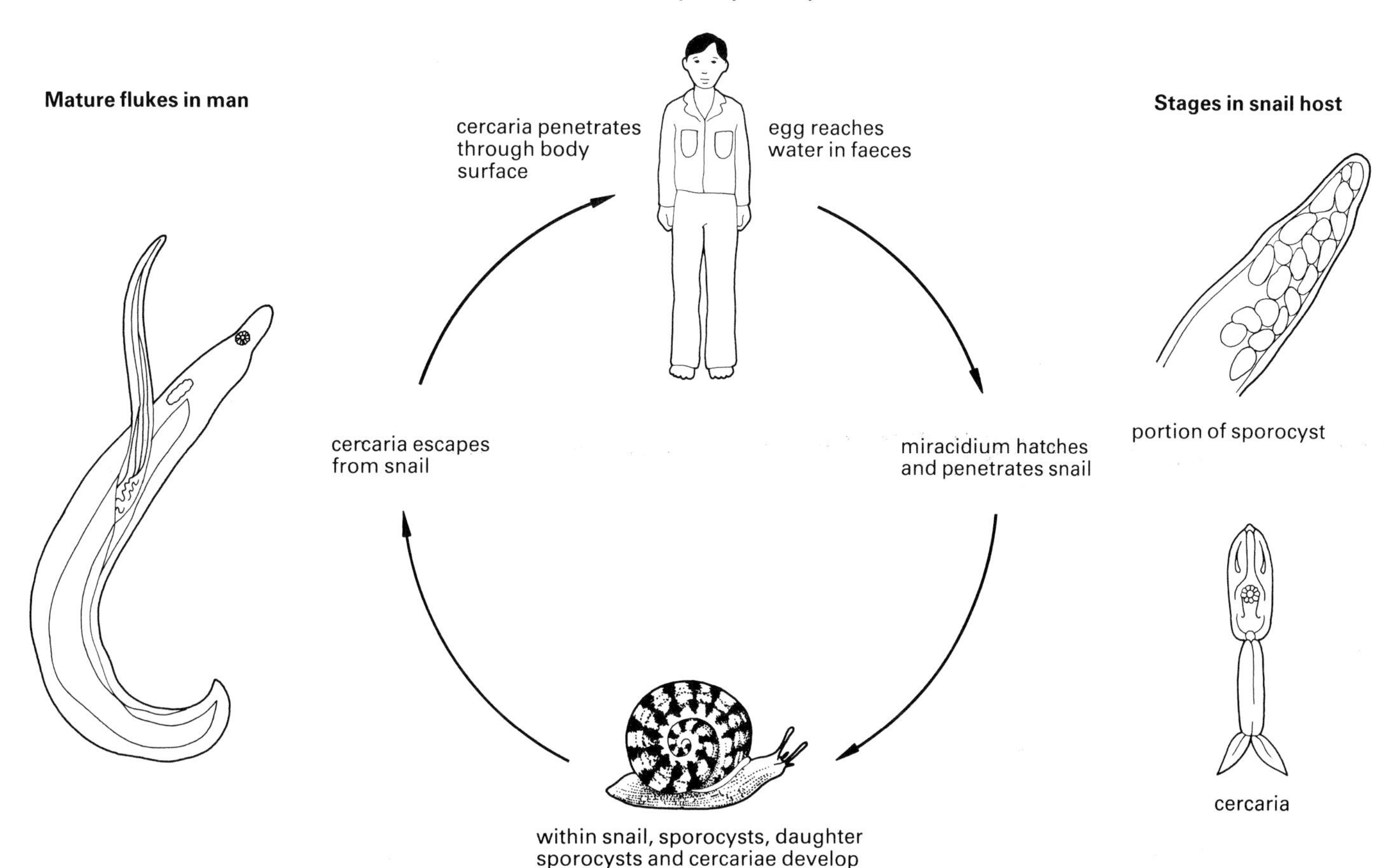

58 The life cycle of *Schistocephalus solidus*

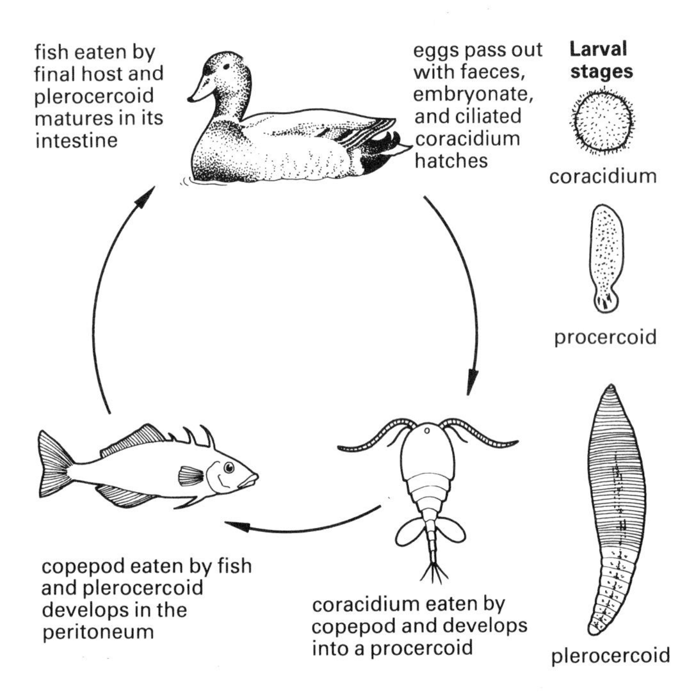

animals necessitates a variety of metabolic and physiological adaptations. It may also induce reactions to the parasite by the host which in turn the parasite must overcome. In vertebrates the defence reactions may be tissue responses or highly specific immune responses (that is involving antigen–antibody reactions).

Animals do not have the monopoly on parasitism either as parasites or hosts. Remember, almost all bacterial, viral and fungal diseases are parasitic and even angiosperm plants can become parasites. Mistletoe is hemiparasitic, its green leaves permitting some photosynthesis, while plants like dodder and *Orobanche* are wholly parasitic on other plants.

Typically one thinks of fungi either as disease-causing parasites, such as athlete's foot fungus and *Piptoporus betulinus* which parasitises and ultimately kills birch trees, or as saprophytic, that is living on dead material. Some fungi are parasitic and then saprophytic. However, the roots of most healthy plants are usually associated with fungi called **mycorrhiza**. This association is so important that many higher plants fail to grow in the absence of mycorrhiza. Characteristically inorganic nutrients

59 Mycorrhiza on a root

60 Mycorrhiza

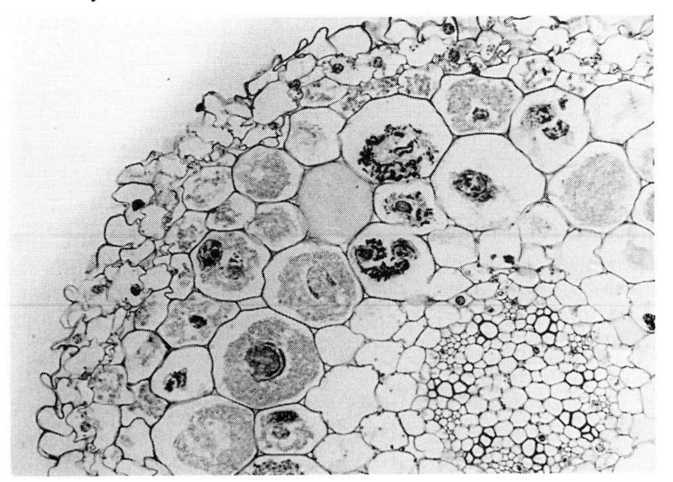

are supplied to the higher plants enabling them to live in low nutrient status soil, and 'in return' the fungus is supported by the plant which is the photosynthesising partner. Mycorrhiza may be endotrophic or ectotrophic, depending on the position of the fungal hyphae relative to the host root; but, more recently, a different scheme of classification has been adopted:

sheathing
vesicular-arbuscular
orchid
ericaceous – ericoid
– arbutoid

The basis of this scheme need not concern us here, but

two of these classes (orchid and ericoid) include a very different type of relationship and are the reason for discussing mycorrhiza in this subsection rather than when we dealt with mutualism.

Certain orchids and members of the group Ericales are without chlorophyll for some or all of their lives. They depend on mycorrhiza to supply all their nutritional needs. In other words these higher plants are parasitising a fungus! *Monotropa hypopitys* is entirely without chlorophyll and dwells in deep forest shade under trees like beech, pine, spruce, fir, oak and cedar. The mycorrhiza connect the *Monotropa* to the tree which is the ultimate source of most of the nutrients for *Monotropa*. In the case of the orchids the relationship is sometimes unstable and environmental conditions may cause the fungus to become parasitic.

Finally, parasitism is not confined to higher organisms. Bacteria may be parasitised by viruses, called **bacteriophages**. Some phages kill their host cells immediately (in which case we might consider the term parasitism inappropriate); others infect host cells which may continue to divide, apparently unaffected, for many generations.

6.4.7 Amensalism

To complete our survey of figure 53, p.51, we have one last category, amensalism. Ecologically it is comparatively uninteresting and covers such activities as accidental damage, trampling and local disturbance of soil nutrients, such as by animals' excretion. It could also be applied to a number of other biotic effects on the habitat, for example shading by trees influencing light and temperature regimes locally.

6.4.8 Concluding remarks

It must be re-emphasised that the purpose of this chapter is to map out areas of study. Accordingly we have taken each of the sections from figure 53, p.51, and discussed something of the types of interaction. Terms have been explained together with the problems of trying to create a rigid classification of types of interaction. You should by now have an overall understanding and be familiar with a range of examples. However, it is possible that you will need to know about certain organisms or associations in greater detail for your examination. Therefore, there is a list of references for further reading at the end of the chapter.

6.5 Interactions between taxa

We have seen how different populations may interact in a variety of ways. We have also seen how some forms of interaction may create selection pressures: the predator–prey relation is a classic example. The prey species develops a defensive chemical or a modified behaviour and the predator responds accordingly. We also saw that this may lead to a progressive specialisation by the predator leading to a dependence on a few or even one prey species.

Whether we are talking of predator–prey or host–parasite relationships or symbionts, there comes a point when to understand the evolution of one we need to understand the story of the other. When evolutionary interactions are interdependent like this we call it **coevolution**. (Coevolution is sometimes used in a more general sense to cover most of the various forms of population interaction).

Becoming an endoparasite (either via ingestion or via the transdermal route) is obviously a complex and lengthy process, and endoparasites that we observe nowadays are the product of millenia of evolution. Once established as a parasite, adaptive radiation occurs; as the host radiates so does the endoparasite. Thus a particular taxon of parasites may be confined to a particular taxon of hosts. The bigger the taxon (species, genus, family and so on) the longer we may assume the radiation has progressed and the earlier the parasitic association first evolved. By way of example, adults of two, highly specialised tapeworm families (Tetrarhynchida and Tetraphyhida) are found only in elasmobranch (cartilaginous) fish.

Researchers have noticed that patterns exist in the food plants of butterflies. Closely related butterflies often feed on closely related groups of plants, for instance the butterfly group Danainae are found on two closely related plant families, the periwinkles (Apocynaceae) and the milkweeds (Asclepiadaceae).

61 The Ehrlich–Raven model of plant–phytophage coevolution

1 Many plant taxa manufacture a prototypical phytochemical that is mildly noxious to phytophages and that may have an autecological or physiological function in the plant

2 Some insect taxa feed upon plants with only this and other, similarly mild, phytochemicals, thus reducing plant fitness

3 Plant mutation and recombination cause novel, more noxious phytochemicals to appear in the plants. The same chemical can appear independently in distantly related plant groups

4 Insect feeding is reduced because of toxic or repellent properties of the novel phytochemical; thus plants with increasingly noxious chemicals are selected for by the pressure of insect herbivory

5 The plant, 'protected from the attacks of phytophagous animals, would in a sense have entered a new adaptive zone. Evolutionary radiation of the plants might follow, ...' (Ehrlich & Raven, 1964)

6 Insects evolve tolerance of, or even attraction to and utilisation of, the novel compound and the plant producing it. An insect can specialise in feeding upon plants with the novel compound; 'here it would be free to diversify largely in the absence of competition from other phytophagous animals' (Ehrlich & Raven, 1964)

7 The cycle may be repeated, resulting in more phytochemicals and further specialisation of insects

A number of observations of this type lead to a model of coevolution between plants and phytophages (plant-eaters) which is summarised in figure 61.

We shall go on to look at just one of the best examples of coevolution between plants and phytophages, but, before we do so, a word of warning. The Erlich–Raven model in figure 61 is highly plausible and suggestive. But, different researchers disagree about how important coevolution is in the mainstream of evolution. That is to say, the intense and reciprocal interactions observed in the following example may be comparatively rare and relatively unimportant, or they may be the stuff of which plant–phytophage evolution is made. This is a current debate.

62 Leaf variation among sympatric species of *Passiflora*. The localities are, from top to bottom, Trinidad, Costa Rica, Costa Rica, Mexico and Texas

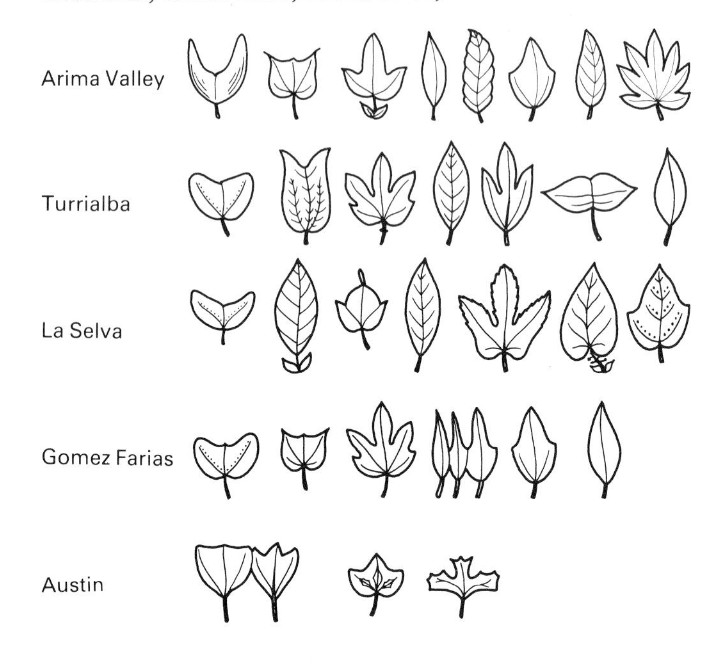

6.5.1 *Heliconius* butterflies and passion vines: an example of coevolution

Heliconius butterflies live in the forests of Central and South America. In their larval stages they feed on the leaves of young shoots of various passion vines, *Passiflora* spp. To reach adulthood they eat so much that the shoot is killed. This is clearly to the detriment of the *Passiflora*. The *Passiflora* produces young shoots regularly and with a patchy distribution (this may be a response to the herbivory by heliconids), furthermore the vines themselves are scattered throughout the forest. The adult female butterfly is obliged to seek out the vines and lay eggs widely distributed and, to enable this, has evolved an energy-efficient flight, large eyes and certain special behaviours. Not surprisingly a butterfly that spends so much time on the wing defends itself by being distasteful and has warning colouration (figure 63).

63 Adult *Heliconius* butterfly

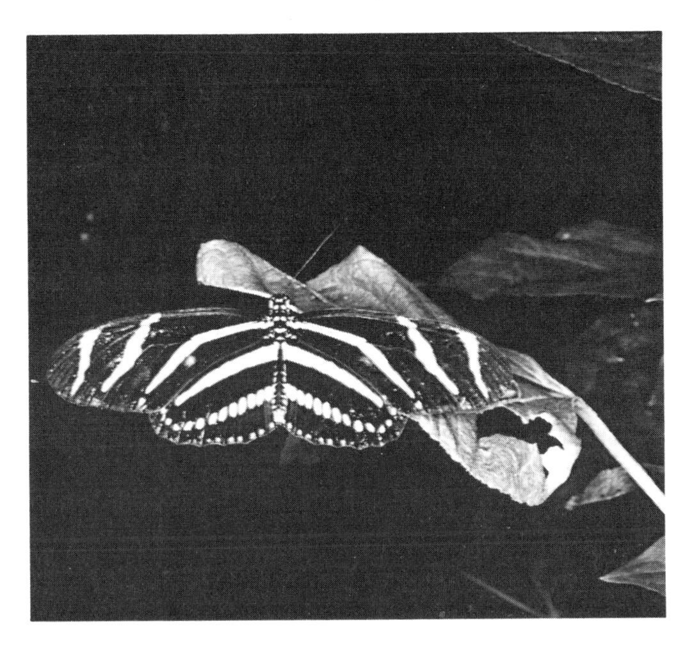

How does the butterfly recognise the *Passiflora?* The answer is presumably by the leaf shape, but the leaf shape varies greatly (see figure 62). Researchers have agreed that this variation is a defence against herbivory: by having a wide range of leaf shapes the butterfly has difficulty in recognising a suitable shoot.

Because of the voracious demands of the larva and its habit of cannibalism, heliconids will only lay one egg per shoot and are reluctant to lay on a shoot where an egg is already present. It looks as if the *Passiflora* takes advantage of this by producing mimetic eggs which deter the heliconid. Stipules, accessory buds, leaf nectaries and petioles have all developed into structures resembling heliconid eggs (figure 64). (Such analogous structures are evidence for strong selective pressure.)

Apart from the *Passiflora* and the heliconids, there is a third component to the system. Adult male and female heliconids require a nitrogenous food, which they obtain from the flowers of another type of vine, *Anguria*. Some *Anguria* species rely on the heliconids for pollination, and it is suggested that the excessive number of male flowers they bear may be an evolutionary response to the need to ensure regular visits.

64 Mimetic *Heliconius* eggs on a stipule of *Passiflora cyanea*. Several days earlier, the new uncurling growth point was sheltered in the stipule so that the two fake eggs were presented in the area where *Heliconius* usually lays its eggs

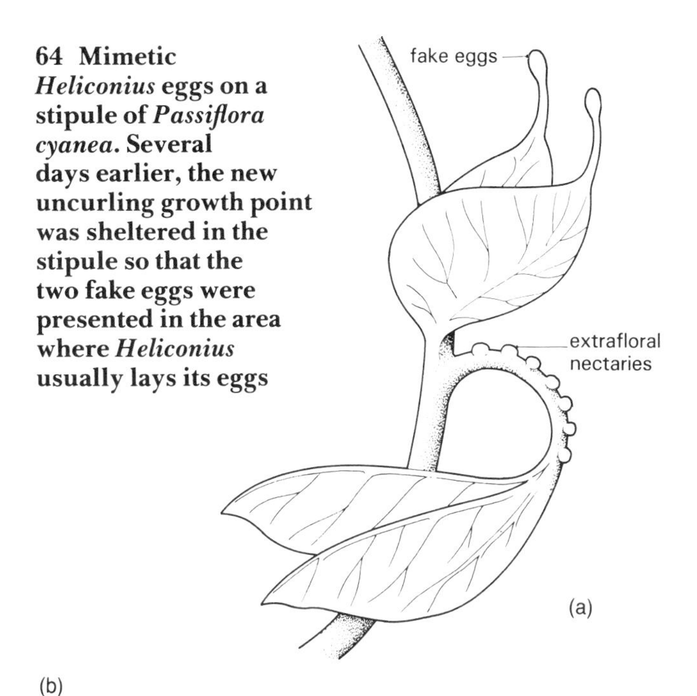

(b)

6.6 References and further reading

Extranuclear Genetics by G. Beale & J. Knowles.
Fungal Parasitism by B. Deverall, Studies in Biology No.17.
The Blue-greens by P. Fay, Studies in Biology No.160.
Mycorrhiza by R.M. Jackson & P.A. Mason, Studies in Biology No. 159.
Plant Symbiosis by G.D. Scott, Studies in Biology No.16.
Symbiosis of Algae with Invertebrates by D.C. Smith, Oxford Biology Readers No. 43.
The Lichen Symbiosis by D.C. Smith, Oxford Biology Readers No. 42.
The Evolution of Eukaryotic Cells by M.A. Tribe, A.J. Morgan & P.A. Whittaker, Studies in Biology No.131.
Sociobiology by E.O. Wilson (the abridged edition).
An Introduction to Parasitology by R.A. Wilson.

There is also a collection of papers from *Scientific American* published as *Animal Societies and Evolution.*

Part II Thematic review

Introduction

The study of biology has traditionally been divided into a number of **topic areas.** These topics deal with certain facets of living things and attempt to present manageable blocks for ease of study. The exact topics vary from syllabus to syllabus. Some common topics are shown in figure 65.

65 Topics in biology

The cell	Reproduction
Tissues and organs	Life cycles
Nutrition	The variety of life
Respiration	Genetics
Transport	Evolution
Homeostasis	Ecology
Behaviour and responses	

A large amount of finely detailed knowledge is needed to fully understand the variety of living forms that exist today and how they interact. The topics in figure 65 present manageable amounts of detail for the initial study of biology. However, a full appreciation of living things can only be gained by looking for **fundamental themes** or **principles** that run across the traditional topic areas and the mass of fine detail. These two aspects of fine detail and basic principle can be illustrated by an example. You may remember that cells are connected by desmasomes (animals) and plasmodesmata (plants) but why should cells be connected at all? Once the knowledge of the names given to these cell connections is gained, it becomes secondary to the more important fundamental principle that cells (organisms also) need to communicate to function efficiently and to survive.

Part II of this unit attempts to highlight some **basic themes** in biology which will require you to draw from your knowledge and understanding of all the topic areas you have been studying. This will help you understand what 'living' means and help you relate the mass of fine detail necessary for A-level examinations into a meaningful framework. Part II should also help you in your preparation for examinations because you will need to review your knowledge and understanding as you work through the unit. The **major themes** presented in this unit and how they relate, **very roughly**, to the traditional biological topics are shown in figure 66.

66 Themes and topics in biology

ORGANISATION	The cell, Tissues and organs, Variety of life, Ecology
CHEMICAL ACTION	Nutrition, Respiration, The cell, Transport, Responses
EQUILIBRIA	Transport, Homeostasis, Responses, Behaviour, Reproduction, Genetics, Ecology
ADAPTATION	Behaviour, Life cycles, Genetics, Evolution
INTERACTION	Ecology

Figure 66 is a vastly simplified summary of part II. When looking for common principles and themes, hard and fast groupings of traditional topic areas cannot be made. All facets of living things interact and overlap. When working through part II, try not to get mentally 'trapped' into traditional topic 'boxes' but look through all the 'boxes' for the **common features.**

Remember, you are looking for the *ways things link up with one another*. Figure 67 shows how the basic themes in biology are interrelated in living systems.

67 Themes and life

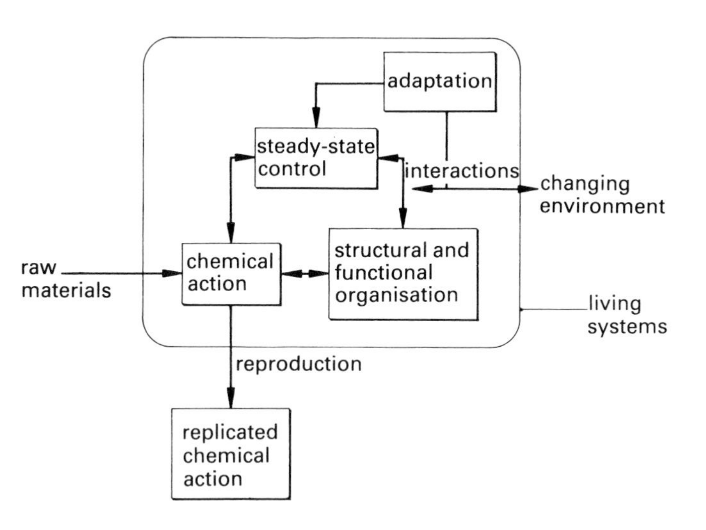

The term **living system** is often used in this unit as many of the themes apply equally to cells, organisms or ecosystems.

The overriding or **prime principle** of all living things is seen as **survival** (the continuation of life processes). All other processes in living things can be viewed in terms of either being advantageous or disadvantageous for survival of the individual or the group. But what do biologists mean by **life?**

You will probably have thought of the **seven characteristics of life:** movement, reproduction, sensitivity, growth, respiration, excretion and nutrition. However, this does not give the whole picture. It is true that individual creatures or **organisms** exhibit particular activities or **functions** when they are alive. However, if these characteristic functions are not being performed, that is the organism is dead, the organism is still distinguishable from **inanimate matter**. Dogs, either living or dead, are organisms, rocks are not. Organisms, therefore, *also* have characteristic organisation or **structures**. Life depends on these structures, which allow life-sustaining functions to occur and to continue (see figure 68).

This part of the unit is not like the other ABAL units. There are no SAQs, summary assignments, self tests and so on. Here, there are only **review**

68 Life defined by structure and function

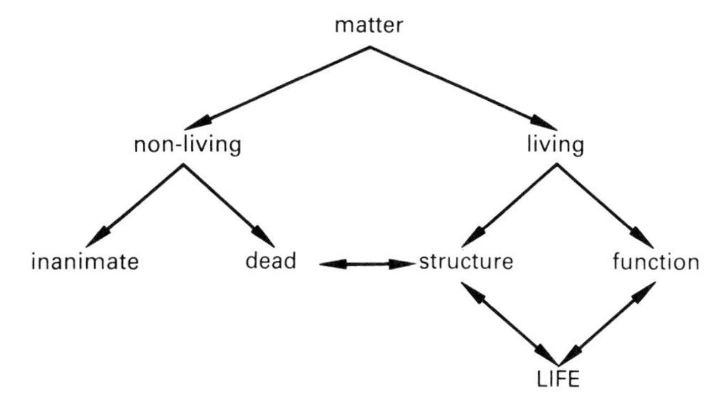

tasks (RTs). Having worked through the other units and periodically looked back over the work you have finished, you should need now to try and piece together all your work into some pattern that has some meaning.

It is hoped that each of the themes shown in figure 69 will form a backbone onto which you can fit the skeleton and flesh of your past studies. The review tasks are *only to indicate* the sorts of information that you should be considering fitting into your pattern at certain points. All the time you should be looking at *all* your knowledge to see what *else* might fit into your skeleton.

69 Themes in biology (the numbers refer to the section in this unit dealing with each theme)

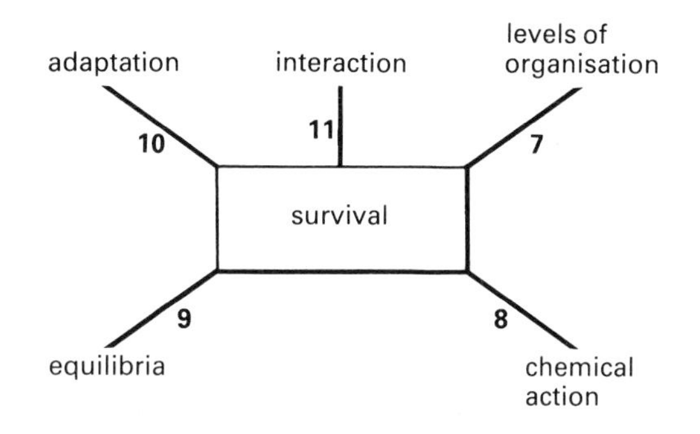

Effective review is a difficult, but worthwhile, task to help you understand the work you have done. Your tutor will give you help in organising your review work.

List of abbreviated review tasks

Levels of organisation

1 Ideas concerning the origin of life
2 Increased complexity and survival
3 (*a*) Chemical structure of organelles
(*b*) Organelle functions
(*c*) Organisation of organelles
(*d*) Diversity of cell structure
4 (*a*) Evolution of early cell types
(*b*) Difference between pro- and eukaryotic cells
5 Cellular requirements for heterotrophic existence
6 (*a*) Classification of unicellular organisms
(*b*) Structure and lifestyles of representatives
7 Colonial organism structure and function
8 Tissue structure and function (brown algae and coelenterates)
9 (*a*) Change in structure and function of plant tissues
(*b*) Changes related to habitat and lifestyles
10 Structure and function of tissues and organs in acoelomates
11 (*a*) Classification of vertebrate tissues
(*b*) Structure and function of each tissue type
12 Structure and function of human organs
13 (*a*) Integration of organs into human organ systems
(*b*) Other vertebrate organ systems
14 The variety of niches and organisms that occupy them
15 Social groupings
16 Structure of communities
17 Types of ecosystem
18 Ecosystem composition
19 Trophic structure and pyramid descriptions
20 Biome locations
21 (*a*) Movement, energy and matter through the biosphere
(*b*) Human activities advantageous and disadvantageous for survival of the biosphere organisation

Chemical action

22 Significance of water for life
23 Properties of buffers and pH
24 Ionic requirements
25 Structure of macromolecule monomers
26 Hydrolysis and condensation reactions
27 (*a*) Molecular functions in living systems
(*b*) Structure of macromolecules
28 Reaction energetics
29 Redox reactions
30 Glycolysis
31 TCA cycle
32 Electron transport system and ATP formation
33 Anaerobic respiration
34 (*a*) Summary of respiratory pathways
(*b*) ATP production
(*c*) Respiratory quotients
35 Photolysis and photophosphorylation
36 Evolution of auto- and heterotrophs
37 Calvin cycle and plant structure
38 Sulphur and nitrogen cycles
39 (*a*) Growth
(*b*) Differentiation and development of cells
40 Secretions of animals and plants and their functions
41 (*a*) Protein synthesis
(*b*) Classification of proteins and enzymes
(*c*) Enzyme action and factors affecting it
42 Intake of raw materials in autotrophs
43 (*a*) Action of digesting enzymes
(*b*) Differences in digestion of different types of heterotroph
44 Diffusion, osmosis and facilitated diffusion
45 Active transport
46 Diffusion and active transport in the functioning of plant and animal organs
47 (*a*) Need for transport systems
(*b*) Functioning of mass flow systems in animals and plants
48 Transduction in animal and plant receptive areas
49 (*a*) Classification of receptor cells
(*b*) Generator potentials
(*c*) Structure and function of sense organs
50 (*a*) Initiation of action potentials
(*b*) Transmission of action potentials
51 (a) Hormonal alteration of DNA activity
(*b*) Muscle contraction
(*c*) Structure and function of effector organs

Equilibria

52 (*a*) Stimulation and inhibition of enzymes
(*b*) Gene regulation
53 Homeostasis in unicellular organisms
54 Homeostatic mechanism in mammals
55 Tolerance
56 Structural and behavioural features aiding homeostasis
57 Immune system
58 Genetic stability
(*a*) Replication
(*b*) Mitosis and meiosis
(*c*) Transfer of genes from parents to offspring
59 (*a*) Mendelian ratios
(*b*) Hardy–Weinberg equilibrium
(*c*) Factors affecting (*a*) and (*b*)
60 Stabilising selection
61 Population dynamics
62 Intraspecific influences maintaining population size
63 Interspecific influences maintaining population size
64 (*a*) Variety of climax communities
(*b*) Forces maintaining the climax community
65 Human effects on ecosystems

Adaptations

66 Meiosis and variation
67 Methods of fertilisation and variation
68 Gene and chromosome mutations
69 Role of environment in characteristic formation
70 Change in gene pools
71 Speciation
72 Theories and evidence for evolution
73 Types of learning
74 Parental care
75 Convergent and divergent evolution
76 Major adaptations in living things

Interaction

77 Embryo development
78 Development of tissues and organs
79 Growth and maturation
80 Internal control and coordination, and biochemical pathways
81 Interactions between individuals
82 Interactions between living organisms and their environment

Section 7 Levels of organisation

7.1 Introduction

Basic self-perpetuating and metabolising matter is made up of cytoplasm and chromatin. It is based upon compounds composed chiefly of carbon, hydrogen and oxygen. All living things are made up of specially structured cytoplasm (except viruses and other virus-like organisms). Amongst the great diversity of living things, a basic theme can be recognised. The cytoplasm of living systems is structured or **organised** according to certain **levels of complexity**. Figure 70 shows the levels of organisation of matter in living systems.

70 Levels of organisation

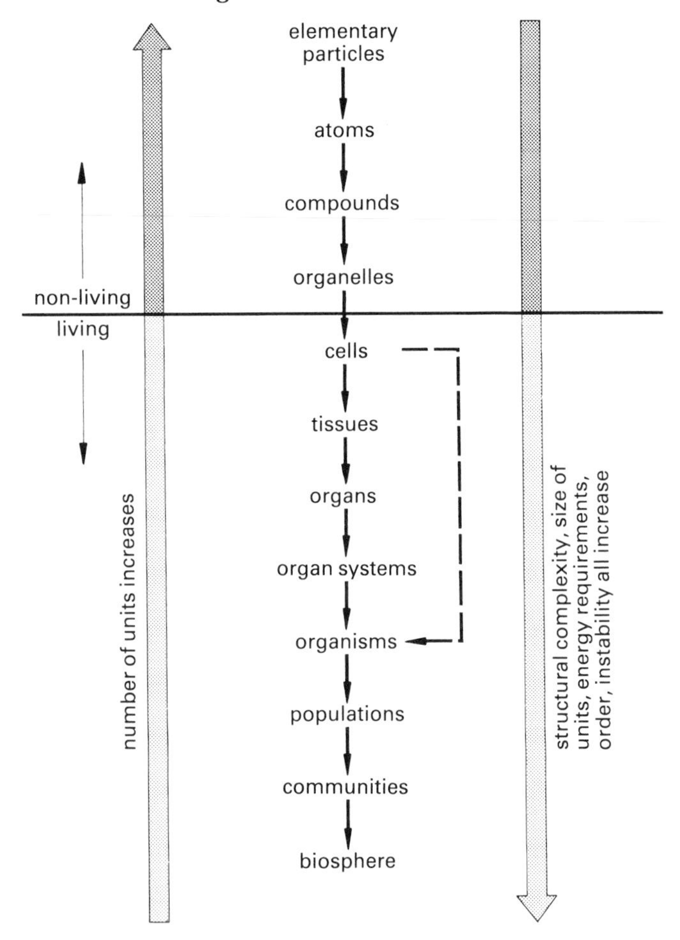

Through evolutionary history, there has been a tendency for living systems to become more and more complex. Figure 70 outlines the past history of life from the non-living origins of chemical compounds to the formation of free-living cellular structures.

RT 1 Review the ideas concerning the origin of life.

From the origin of the first life forms, organisms have become progressively more complex. Each level of increased complexity has its own particular properties that are more advantageous for the survival of particular lifestyles of organisms.

RT 2 How is increased complexity of value in survival?

Remember, organisms exist at each of the organisational levels shown. They also have the organisation complexities of all the levels lower than their own. For example, unicellular organisms have atomic and molecular organisation. Organisms that have developed tissue systems have tissue organisation *as well* as the atomic and molecular organisation of those lower levels.

Note: in this section, particular organisms will be mentioned that have *reached* certain levels of organisation and their structures considered. More highly organised organisms will also *show* this level of organisation but the details of their structures are dealt with at the maximum level of organisation reached by the organism (see figure 71).

7.2 Cellular organisation

Compounds such as proteins, carbohydrates and lipids form complexes in living systems. Such complexes are called **organelles**. Individual organelles are *not* living, only when organelles are combined in a specific manner are life processes possible. This specific combination of organelles is the unit of life, **the cell.**

71 Levels reached and levels shown by organisms

Levels of organisation	Level reached			Levels shown		
	Paramecium	*Hydra*	Human	*Paramecium*	*Hydra*	Human
Cell	✓			✓	✓	✓
Tissue		✓			✓	✓
Organ						✓
Organ system			✓			✓

RT 3 (*a*) Describe how macromolecules are organised to form the following organelle structures: plasma membrane, endoplasmic reticulum, ribosomes, mitochondria, nucleus, lysosomes, cilia and flagella, centrioles, Golgi body, chloroplasts, cell walls.
(*b*) Describe the functions performed by the organelles.
(*c*) Describe how the organelles are organised to confer life by forming typical animal and plant cells.
(*d*) Review the diversity of cellular structure formed by modification to this basic organisation of organelles.

During the evolution of life, the very earliest cells developed into a variety of basic cell types still in existence today.

RT 4 (*a*) Review the evolution of the earliest cells.
(*b*) Outline the differences between **prokaryotic** and **eukaryotic** cells.

Note that even at this relatively low level of organisation the structure of life is of great importance. The special organisation of the cell determines the lifestyle of the individual. A cell lacking chloroplasts almost always adopts a heterotrophic existence (there are very few organisms without chloroplasts that are autotrophic). The absence of this one organelle has far-reaching consequences. A heterotrophic unicellular organism must acquire further cellular specialisations that are required for this particular lifestyle.

RT 5 Review the cellular requirements for a heterotrophic existence.

If a single cell is capable of performing *all* the functions of living systems, it is considered to be an individual **organism**. Many such unicellular organisms have a surprisingly elaborate organisation associated with their cytoplasm and nucleus.

RT 6 (*a*) Review the classification of unicellular organisms.
(*b*) Give details of the structure and lifestyles of representatives from each group.

7.3 Multicellular organisation

In the majority of organisms, life processes are carried out by more than one cell and such organisms are termed **multicellular.** Different levels of organisation occur depending upon the number, types and associations of cells found within such organisms.

7.3.1 Colonial level

The simplest multicellular level of organisation occurs when single-celled individuals become attached together with little or no communication. The cells of such a **colony** may show specialisations of structure and, therefore, function. Relatively few cells are involved at this level of organisation.

This limited specialisation of structure and function and lack of communication between cells results in little **division of labour** within the organism. An organism is anything that performs all the living processes. When different cells become specialised and 'share out' the functions of the organism as a whole, division of labour is taking place.

RT 7 Give examples of colonial organisms and describe their structure and function.

7.3.2 Tissue level

Organisms composed of large numbers of cells invariably have the cells organised into **tissues,** regions of cells **specialised** and **coordinated** to perform a particular function. Different organisation allows an increase in specialisation of cells and increased division of labour, but this also brings increased **interdependence** of cells within the organism. Food-absorbing cells become dependent on other cells for protection and, therefore, survival (and vice versa).

Examples of organisms that are organised at this level of complexity are seen in the brown algae and the coelenterates.

RT 8 Describe the structure of tissues and their functions found in organisms that have reached the tissue level of organisation.

In simple multicellular animals there are typically two cell layers, the **ectoderm** and **endoderm**, often with a jelly-like mesogloea in between. Animals that have these cell layers developed into tissues are called **diploblastic**. (Multicellular organisms without tissues, the colonial forms, are not usually referred to as diploblastic.)

7.3.3 Tissue systems

This level of organisation mainly concerns the elaboration of tissues and their **interrelation** to form **tissue systems**. The majority of plants have reached this level of organisation. Because of their relatively 'simple' mode of nutrition, plants have reached the level of complexity appropriate to their lifestyles.

Survival has been ensured by modification of the tissues to meet the dangers and opportunities that occur. The major increase in complexity of plant tissues and formation of tissue systems have come about as a result of the colonisation of a variety of habitats.

RT 9 (*a*) Review the changes in structure and function that have occurred in plant tissues from those seen in the brown algae.
(*b*) Relate these changes to the habitats colonised and the lifestyles of named plants.
(Pay special attention to the tissues involved in water and ion uptake, anchorage, support, protection, gas exchange, photosynthesis, transport of substances and reproduction.)

7.3.4 Organ level

Many animal groups have structures where a number of different tissues have become associated and coordinated to perform a particular major function of the organism. Such structures are called **organs**. There is some debate as to whether roots, leaves and so on, of plants constitute organs, such as the leaf as an organ of photosynthesis. This unit assumes plants do not contain true organs but specialised tissue systems.

Animal groups that develop organs are also **triploblastic** (have three tissue layers), the extra **mesoderm** replacing the mesogloea and contributing largely to organ formation. Animals that have reached the organ level of organisation can be divided into two groups depending upon whether a body cavity develops within the mesoderm layers or not (**coelomates** and **acoelomates** respectively) (see figure 72).

72 Development of the coelom

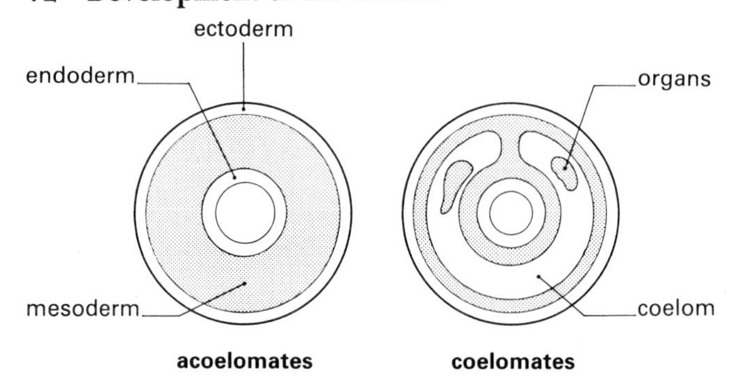

The presence of organs and more specialised tissues in general requires the development of greater controlling and coordinating systems in the organism for efficient functioning.

RT 10 Describe the structure and function of tissues and organs found in planarians (acoelomates).

7.3.5 Organ systems level

Just as cells and tissues operate more efficiently when coordinated with one another, organs too are interrelated to form **organ systems**. In this level of organisation, a number of different organs are coordinated into an organ system which achieves a number of related functions aimed at efficient performance of a major activity of the animal. Organ systems are found in vertebrates; examples include digestive, urinogenital, musculoskeletal, immune, respiratory, vascular, sensory.

Before looking at the overall structure and functions of these organ systems, it is necessary to look at the structure and function of the constituent vertebrate tissues and organs. Their organisation and complexity are relatively simple when compared to those of organisms at the organ systems level of organisation.

RT 11 (*a*) Review the classification of vertebrate tissues (including muscle, blood and nerve tissues).
(*b*) Outline the structure and function of a variety of tissues from each tissue type.

These basic tissues are combined in a variety of ways to form organs. Human animals are most often considered when studying vertebrate organs.

RT 12 Review the structure and function of the following human organs, remembering to highlight the tissues involved in their construction: lung, liver, intestine wall, heart, pancreas, brain and spine, kidney, ear, gonads, lymph glands, skin, eye.

RT 13 (*a*) Review the way in which the above organs are structurally interrelated to form the organ systems listed above.
(*b*) Review the organ systems of any other vertebrate you have studied (for example fish).

7.4 Group organisation

Sections 7.2 and 7.3 have been concerned with the organisation of individual organisms and how specialisations of cells and their association have contributed to increasing the efficiency of living systems. Individual organisms, like cells or organs, associate to form efficiently functioning living systems. Over evolutionary history, different groups have developed different and specialised lifestyles that interrelate the organisms with their physical and biological surroundings. This section deals with the level of organisation where individuals are organised into groups of varying sizes and complexities. (Section 11 considers the interactions taking place within and between these living systems.)

7.4.1 Species

A **species** is a group of very similar individuals that have in common certain structural, functional and evolutionary traits. A member of a species is normally only capable of breeding (to produce fertile offspring) with members of the same species. Due to their structural and functional characteristics, each species also has its own particular **mode of existence**, its role in nature or **ecological niche**. Nature offers opportunities for terrestrial, aquatic and aerial niches. Within each of these there are arctic, temperate and tropical niches and day- and night-time niches and so on.

RT 14 Review the variety of niches and give examples of species that occupy them.

Survival of the species depends on how well fitted (adapted) it is to its niche, how well it can compete with other species for a particular niche and how well it can adapt to other niches. These adaptations and interactions are considered in sections 10 and 11.

7.4.2 Populations

A species is a group of organisms that are **capable** of interbreeding. Members of a species often do not interbreed due to physical barriers that keep organisms apart. These barriers result in smaller groups of individuals of the same species geographically or spatially localised that **actually** interbreed. Such a group is called a **population.** For example, the common oak is a species found throughout England. A localised breeding group, such as a forest, will probably not interbreed with another oak forest a few miles away. The common oak species is, therefore, split into a number of populations throughout the country.

The individuals of a population are organised as a group due to their similar **environmental requirements** which are geographically or spatially localised. In many plant and animal populations, the members of the group live mainly as solitary individuals and, apart from reproductive associations, they show relatively few other interactions.

However, in some animal groups, there is extensive and direct interaction between individuals. Such a social grouping of **society** has many advantages for survival.

RT 15 Review the structure and benefits of a variety of social groupings.

7.4.3 Communities

A particular locality usually contains more than one population. Such a group of populations from different species is called a **community**. Almost always, a community includes interdependent associations of plants, animals and microbes. Like individual organisms, the structure of a community depends upon its age. A community grows and develops with time through the process called **succession.**

RT 16 Outline the structure of the following communities: oak wood, sand dune, pond and any other community you have studied.

7.4.4 Ecosystems

All the communities and their physical surroundings that exist in a particular, discrete area are termed an **ecosystem.**

RT 17 Review the variety of types of ecosystems that exist.

Although there are many types of ecosystem, just as there are many types of mammal, all ecosystems have the same basic organisation of structures and functions (see figure 73).

The exact numbers and types of the populations that constitute the producers and consumers and so on,

73 Organisation of ecosystems

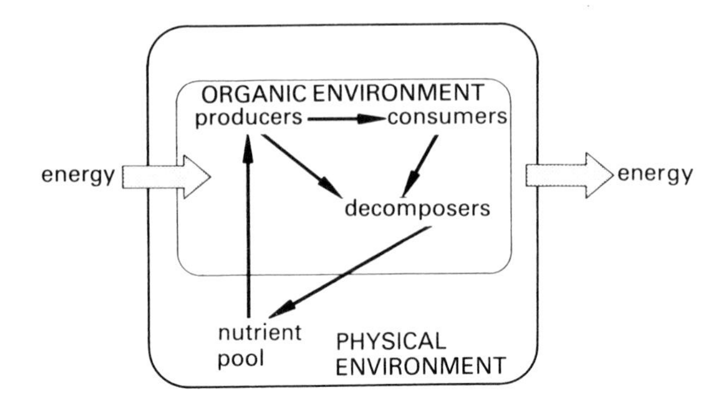

in a particular ecosystem will depend largely on the physical surroundingings.

RT 18 Review the exact composition of a number of ecosystems, each of which is exposed to different physical surrounds.

Describing the structure of ecosystems in terms of the variety of populations that can be present, emphasises the diversity of ecosystems. However, the basic similarity of many ecosystems can be shown by describing the fundamental functions of energy flow through, and the matter cycling within, ecosystems.

RT 19 Review the basic trophic structure of ecosystems and the pyramid descriptions that can be applied.

7.4.5 Biomes

On a larger scale, the organisation of similar ecosystems into a grouping constitutes a **biome**. Biomes are really zones of major vegetation types, for example arctic tundra, tropical rain forests, deserts. Latitude and longitude mainly govern their structure and distribution. Biomes differ in the structure of the vegetation, the types of flora and fauna, geographical ranges and climatic regimes.

RT 20 Outline the major biomes and their location on the Earth.

7.4.6 Biosphere

All biomes together and the volume of the Earth where living things are found make up the

biosphere, the top of the structural organisation of living matter (see figure 74).

Like an individual organism, the living planet is made up of a complex organisation of smaller, living systems that have particular structures and functions. For the biosphere (and, therefore, all life) to survive, all its components must function efficiently and in balance.

74 The biosphere

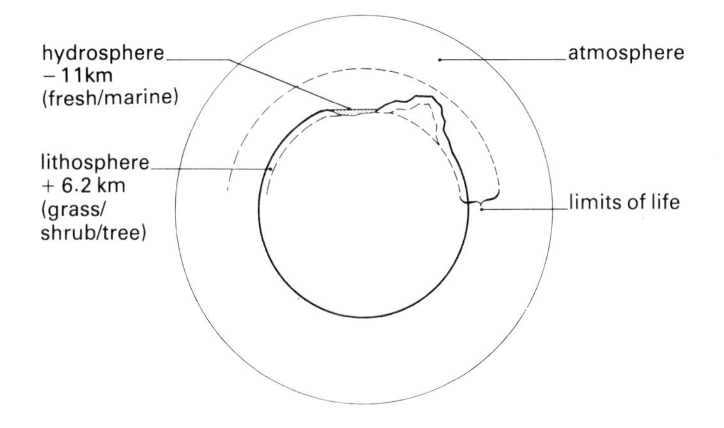

RT 21 (*a*) Review the movement of energy and matter through the biosphere.
(*b*) Review the activities of human populations that are disadvantageous and advantageous for the survival of the organisations of the biosphere.

This section has considered the characteristic organisations of living systems from the cell to the biosphere. It is not really possible to consider the structure and function of living things without mentioning how they interact with one another. This theme will be considered further in section 11.

Section 8 Chemical action

8.1 Introduction

Having looked, in section 7, at the characteristic structure of living things, this section considers one of the characteristic functions of life: the acquisition and treatment of raw materials and energy (these processes are collectively called **metabolism**). Metabolic processes of living things are exclusively performed by the coordinated action of a great variety of chemicals.

This section looks at the common processes performed by organisms, the chemicals concerned and the functions they perform. Chemical action in living things can be summarised under three broad headings:

(*a*) the liberation of energy from the environment for use by the organism for further chemical action,
(*b*) to make available raw materials from the environment for building body substances,
(*c*) to initiate or accomplish a variety of other processes.

Figure 75 summarises metabolic activities of living things.

75 Functions of chemical action

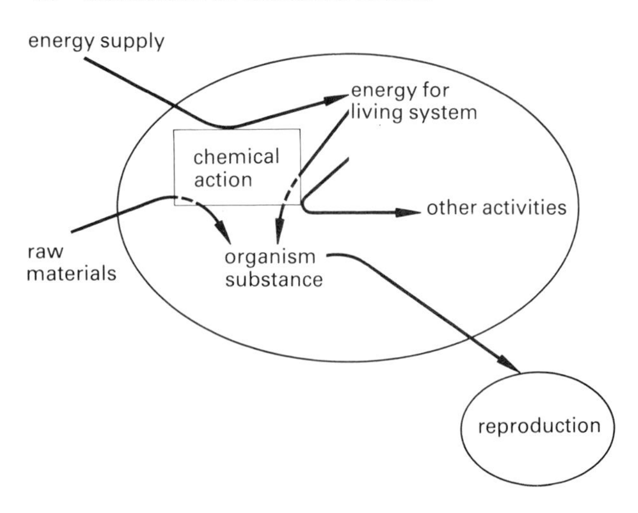

8.2 Chemicals of life

Section 7.2 considered how complexes of proteins, carbohydrates and lipids formed the organelles of cells. In this section, you will consider the structure of the molecules that make up living matter.

8.2.1 Inorganic compounds

Approximately 70% of the cell is composed of water. It is the medium for chemical action in living things and so the properties of water are of great importance for life.

RT 22 Review the properties of water with reference to their significance for living things.

Biochemical reactions require steady values of acidity (pH) for their correct operation.

RT 23 Review the action of acids, bases and buffers and explain how their reactions affect the pH of solutions.

Other important inorganic constitutents of living things are **mineral salts,** ionic compounds containing a metallic and a non-metallic ion (see figure 76).

76 Ions of mineral salts

Metallic ions (cations)	Na^+	K^+	Ca^{2+}	Mg^{2+}	Cu^{2+}	Fe^{2+}
Non-metallic ions (anions)	Cl^-	HCO_3^-	NO_3^-	$H_2PO_4^-$	SO_4^{2-}	I^-

RT 24 Review the structures and processes in animals and plants that each of the ions shown in figure 76 are involved in.

8.2.2 Organic compounds

The principal organic compounds mentioned previously are all large molecules or

macromolecules made up by repeating organic units or **monomers**. In each class of macromolecule (carbohydrate, lipid, protein and nucleic acid) the monomers are different.

RT 25 Review the structure of the basic units from which the four classes of macromolecule are composed.

Despite the differences in organic monomers and the actual type macromolecule produced, the reactions involved in building up or breaking down these macromolecules are very similar.

RT 26 Review the **condensation** and **hydrolysis** reactions and the bonds formed during synthesis and degradation of the four types of macromolecule.

The importance of these molecules to living things lies in the functions they perform. These **functions** can be broadly divided into **energy supply, structural, informational, catalytic** and **accessory**. This division according to functional similarity is probably of more use during revision than a division according to chemical similarity. For example, most carbohydrates are concerned with energy supply but some are of great importance in the structure of plants.

RT 27 (*a*) Using the above-named functional categories as headings, review the variety of named molecules and their particular function in living systems.

Note: A group of organic compounds you should include in this review are the additional food factors required in many metabolic processes (that is the **vitamins**).

(*b*) Review the structure of the molecules mentioned in 27 (*a*).

So far, you have looked at the structure and function of chemicals found in living systems but not at how these molecules are built, or **synthesised**, by organisms. This synthesis (and all other processes) requires the living system to expend energy. The most fundamental chemical actions are those that liberate energy for use by the organism (section 8.3). Section 8.4 considers the use of this energy in the synthesis of the chemicals of life.

8.3 Energy liberation

All matter contains energy. When atoms bond together to form compounds, molecules and so on, more energy is incorporated into the molecules. Energy liberation in living things involves breaking the chemical bonds and **conserving** the energy by forming bonds in other molecules (such as ATP) directly usable by the organism's chemical system.

Chemical energy of the metabolites is released mainly by breaking the bonds between carbon and hydrogen and combining these atoms with oxygen (see figure 77).

77 Energy liberation in living things

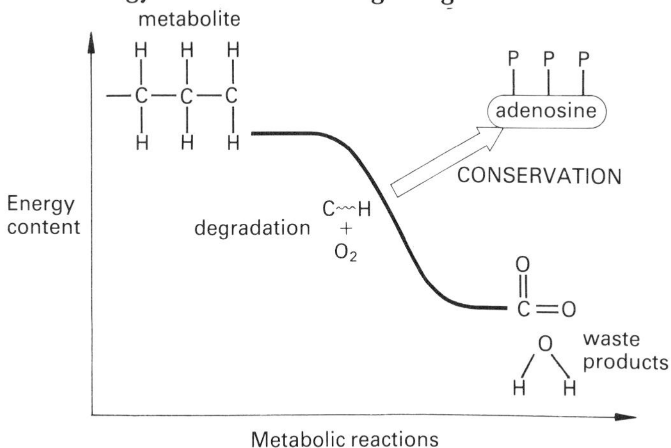

RT 28 Review the energetics of **exothermic** and **endothermic** reactions.

Reactions that involve the movement of hydrogen, oxygen and electrons are termed reduction and oxidation reactions or **redox reactions.**

RT 29 Review processes involved in redox reactions.

Although energy liberation is often termed oxidation because the overall reaction shows the addition of oxygen to the metabolite (see figure 78) the actual chemical process involves the removal of hydrogen **(dehydrogenation).**

78 Overall reaction for glucose

$$C_6H_{12}O_6 + 6O_2 \longrightarrow 6CO_2 + 6H_2O$$

In the decomposition of metabolites for energy liberation it is better to think in terms of **dehydrogenation**, as the addition of oxygen (if it actually occurs) takes place at the very end of the biochemical pathways (see figure 79).

79 Energy liberation by dehydrogenation

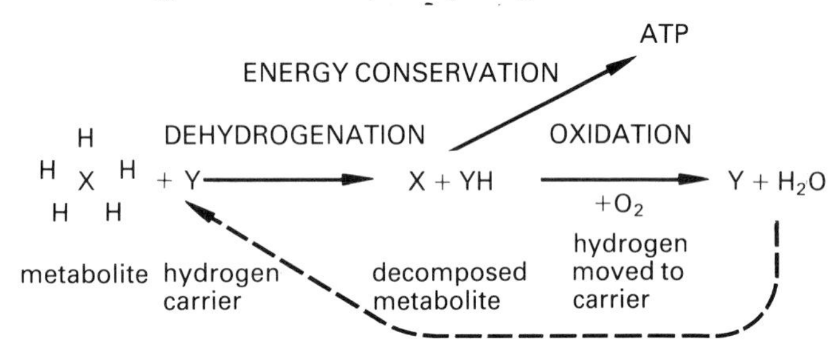

Every living cell provided with adequate supplies of raw materials (metabolites) can liberate the chemical energy of the metabolites and conserve it in the form of adenosine triphosphate molecules (ATP). ATP is the main form of chemical energy usable in the metabolic processes of living things.

ATP can be generated in two main ways. Firstly, phosphate can be added to energy metabolites and then transferred to ADP during molecular rearrangements **(glycolysis)**. Secondly, electrons liberated from dehydrogenated (oxidised) metabolites can initiate ATP formation by an electron transport system **(oxidative phosphorylation).**

All cells can carry out these rearrangement and oxidative reactions that form ATP. In addition, photosynthetic cells can convert light energy into the chemical energy of ATP molecules by a process of **photophosphorylation.**

8.3.1 Rearrangement and dehydrogenation reactions

Figure 80 shows the generation of ATP by the rearrangement reactions (glycolysis) involving a six-carbon metabolite. It is the sequence of dehydrogenation and rearrangement reactions that generate ATP.

RT 30 Review the details of the chemical reactions involved in glycolysis.

80 ATP generation during glycolysis

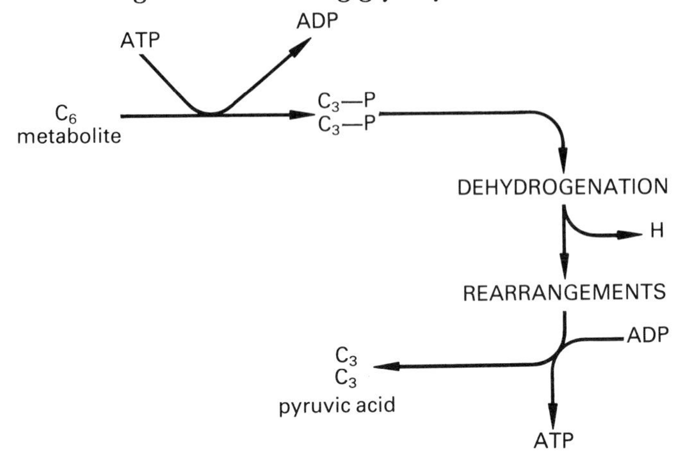

Very little ATP is generated for the organism by the series of reactions in glycolysis. Further dehydrogenation of the metabolite, now in the form of two three-carbon compounds (pyruvic acid) does not generate ATP directly (see figure 81).

81 Dehydrogenation of pyruvic acid

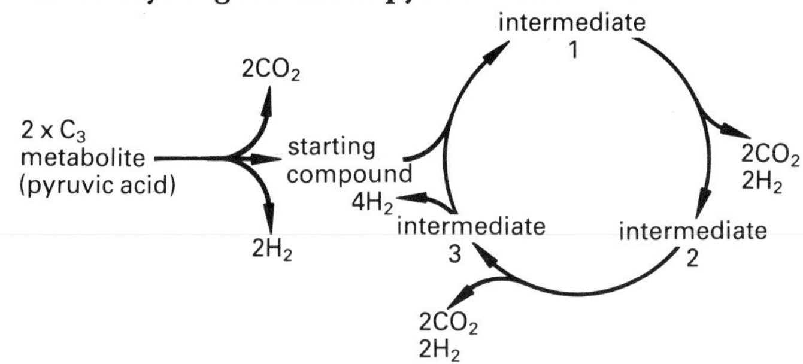

RT 31 Review the details of the chemical interconversions involved in the **Krebs (TCA) cycle.**

The importance of these oxidation/hydrogenation reactions is the liberation of hydrogen which is then used by the accessory **electron transport system** for the generation of ATP (see section 8.3.2).

8.3.2 Oxidative phosphorylation

Energy metabolites (most often glucose, more rarely others such as H_2S,) are mainly a source of hydrogen atoms. It is when the hydrogen atoms give up electrons that the main amount of energy is liberated and ATP is formed (see figure 82).

RT 32 Review the redox reactions occurring in the electron transport system and ATP formation.

82 Utilisation of hydrogen in aerobic conditions

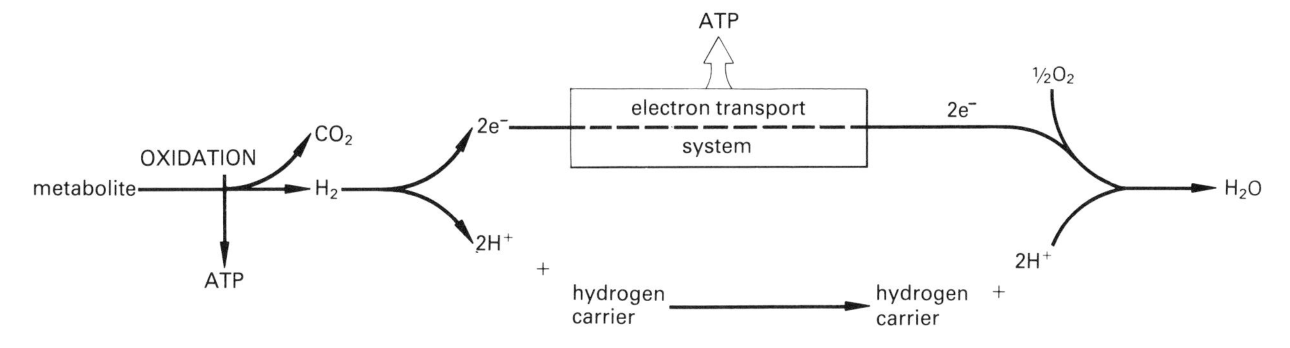

83 Reactions of hydrogen in anaerobic conditions

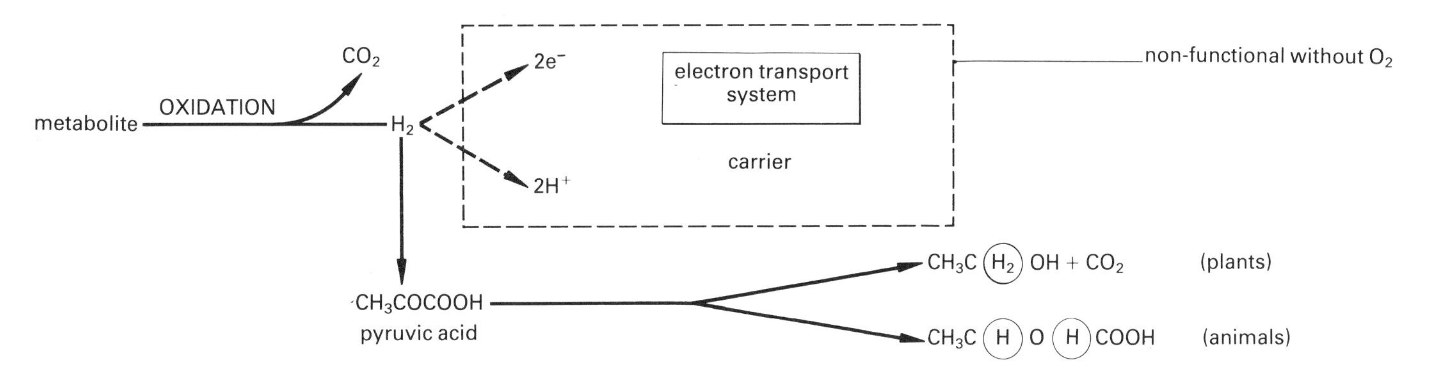

84 Incorporation of organic compounds into energy-liberating reactions

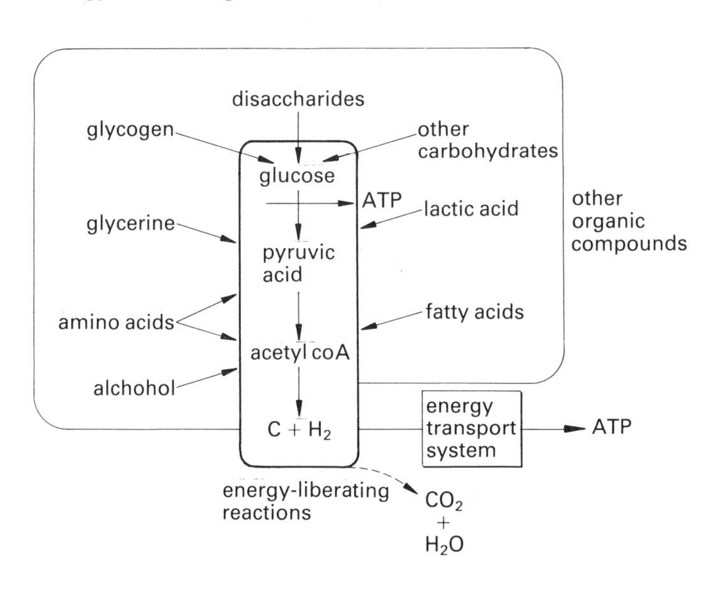

This series of redox reactions shown in figure 82 occur in the walls of mitochondria of both animal *and* plant cells. The reactions will not operate though unless the hydrogen ions and electrons are removed by reaction with oxygen **(aerobic)** at the end of the chain.

Many cells, however, can function for short periods without oxygen **(anaerobically)**. In this situation, hydrogen atoms liberated by dehydrogenation in glycolysis and the Krebs cycle are accepted by pyruvic acid (see figure 83).

RT 33 Review anaerobic respiration and its significance in a variety of organisms.

Any organic compound can be decomposed to carbon and hydrogen by conversion of the compound into intermediates of the above energy-liberating reactions. In times of food shortage, almost any part of the organism can be used as an energy source (see figure 84).

The term used to describe the liberation of energy from metabolites by these rearrangement, dehydrogenation and phosphorylation reactions is **respiration.**

RT 34 *(a)* Review the chemical pathways of respiration with attention to the generation of carbon dioxide, hydrogen (electrons) and ATP molecules with glucose as the metabolite.
(b) How do the number of ATP molecules generated differ for different substrates?
(c) How does the use of oxygen and the production of carbon dioxide **(respiratory quotient)** compare for different substrates?

8.3.3 Photophosphorylation

Cells of plants perform the series of reactions outlined in 8.3.1 and 8.3.2. However, photosynthetic cells also contain energy-liberating organelles other than mitochondria; these are the **chloroplasts**. Chloroplasts can use light energy to form ATP molecules by a process called **photophosphorylation.**

Light energy causes chlorophyll molecules to lose electrons which can pass along the electron transport system and form ATP molecules. Some electrons return to chlorophyll, but many are removed by carriers to be used in other reactions (see section 8.4.2) before they can enter the electron transport system (see figure 85).

85 Electron transport and ATP production in chloroplasts

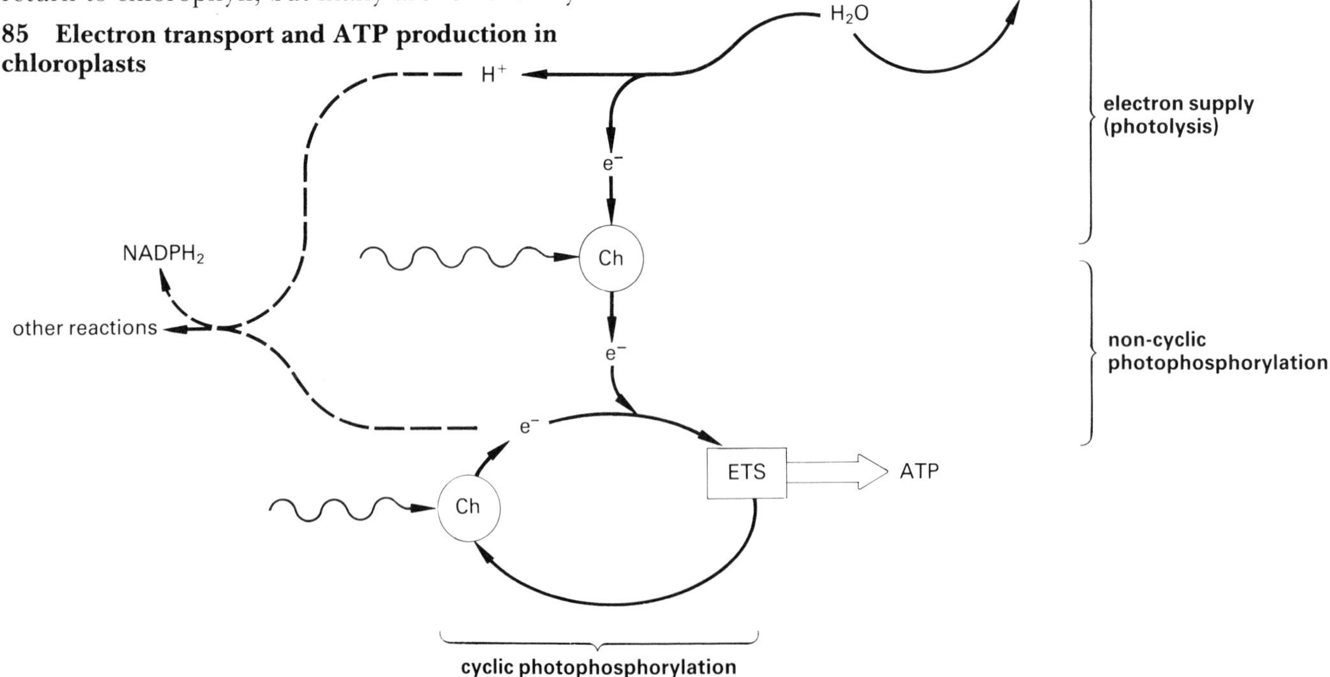

86 Cyclic and non-cyclic photophosphorylation

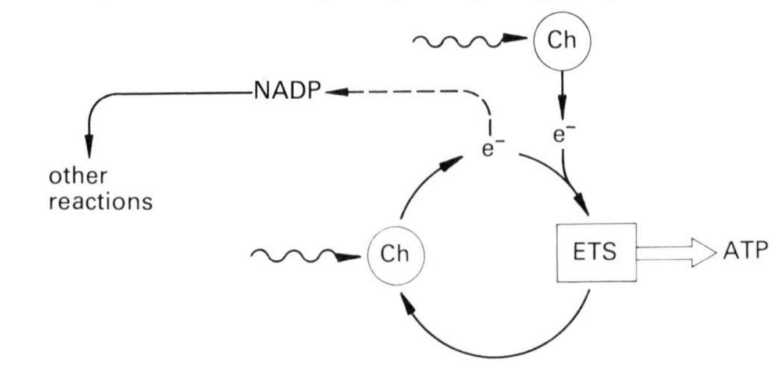

light energy
Ch chlorophyll molecules
e^- electrons
ETS electron transport system

Without a supply of electrons, the chlorophyll would quickly stop functioning. **Water** molecules act as the electron supply for chlorophyll. By looking at the movement of electrons, it can be seen that there are two types of photophosphorylation, **cyclic** and **non-cyclic** (see figure 86).

RT 35 Review the details of photolysis, cyclic and non-cyclic photophosphorylation.

The amount of ATP derived from photophosphorylation is insufficient to sustain life in plants and must be supplemented by ATP from oxidative phosphorylation in mitochondria (or glycolysis in anaerobes).

The main significance, however, of these **light-dependent reactions** is the generation of hydrogen ions and electrons for reactions that involve the **reduction of carbon compounds** (section 8.4.2).

8.4 Matter availability

8.4.1 Introduction

Although all living things are composed mainly of protein, carbohydrate and lipid, every individual has unique chemical characteristics. The types of protein carbohydrate and lipid and the way they are arranged in every cell of the organism are different even in closely related individuals. This is a fundamental process involved in the **immune response** (see section 9.3.2).

This uniqueness is as a result of every organism **transforming**, or making available, the matter it takes in before it can produce its own **characteristic body substance.** This section looks at how organisms acquire matter, degrade it to its constituent parts and synthesise their own body substances.

8.4.2 Synthesis of body substances

Early in the evolutionary history, many organic compounds were formed by physical phenomena in the primeval atmosphere. Once living cells become established (section 7.1) these organic compounds rapidly become incorporated into the living things. As this organic soup became depleted, some organisms developed chemical processes to synthesise their own organic compounds from the abundant inorganic compounds available in the environment. Such **autotrophic** organisms are the ultimate source of all organic compounds that act as building-blocks for all living things.

RT 36 Review the early evolution and divergence of autotrophic and heterotrophic organisms.

87 Synthesis of the basic carbohydrate PGA

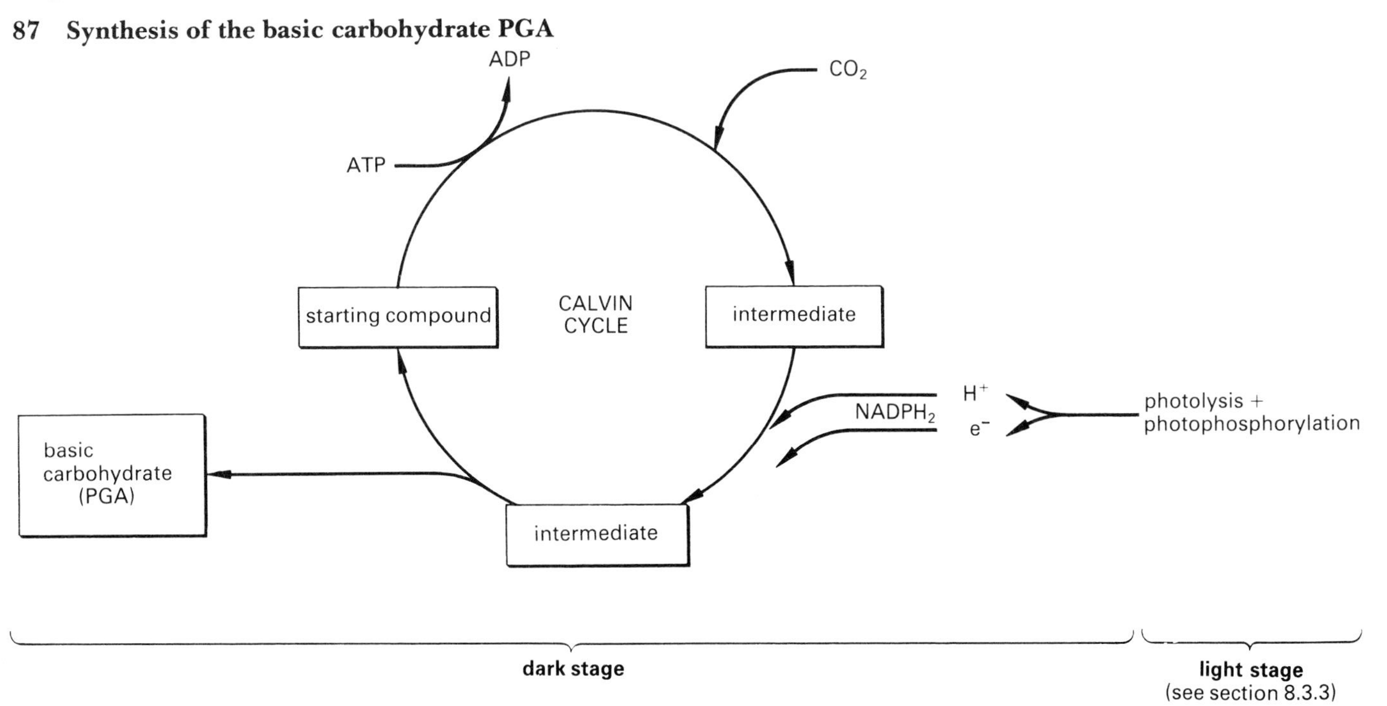

Given energy and raw materials, chloroplasts manufacture the basic carbohydrate **phosphoglyceraldehyde** (PGA) in a series of reactions called the **Calvin cycle** (see figure 87).

It is photophosphorylation (light stage) and the Calvin cycle (dark stage) that constitutes the process called **photosynthesis.**

RT 37 Review the details of carbon fixation in the Calvin cycle and relate chemical action during photosynthesis to structures in plants.

Living things are made up of more than just carbohydrate, and so autotrophs must be able to synthesise the other building-blocks: amino acids, fatty acids, glycerol and nucleotides. Figure 88 shows a summary of the synthetic pathways by which the organic compounds basic to life can be built up.

88 Organic synthesis

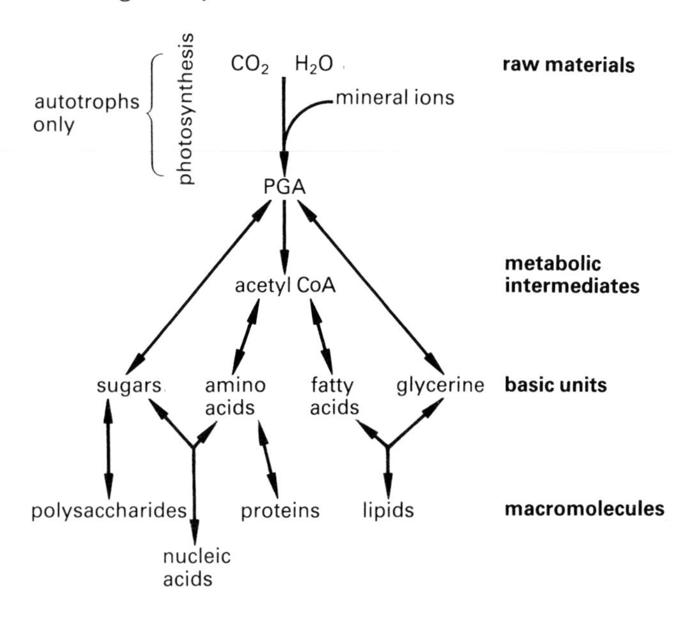

Extremely important in the synthesis of organic compounds are the **autotrophic bacteria**. During their metabolic reactions, some types of bacteria make available chemicals such as sulphur and nitrogen in a form that is usable by plants (see figure 89).

89 The importance of autotrophic bacteria

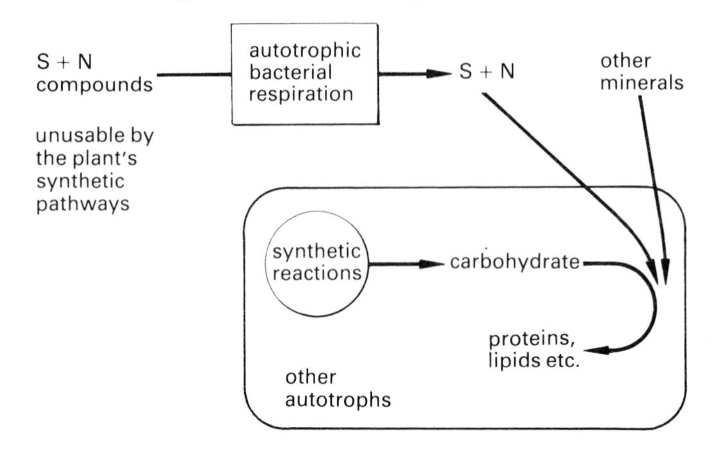

RT 38 Review the details of the chemical interconversions involved in the sulphur and nitrogen cycles.

The organic pathways shown in figure 88, with the exception of PGA synthesis, also occur in animals and fungi (heterotrophs). The building of body substances, therefore, is very similar in animals and plants although the emphasis of synthesis is different. (Mainly carbohydrate in plants and protein in animals).

The products of synthesis are needed for replacement of molecules within cells and for the growth and development of the organism. Cell division produces cells that must synthesise all the macromolecules needed for growth, further division and differentiation.

RT 39 (*a*) Review the processes involved in growth in animals and plants.
(*b*) Review the differentiation and development of a variety of cells with emphasis on the synthesis of new body substances.

An important role of organic synthesis is in the formation of secretions for the proper internal and external working of animals and plants.

RT 40 Review the variety of secretions produced by animals and plants and give their functions.

Perhaps one of the most important group of body substances to be synthesised is the catalytic proteins or **enzymes.** Practically all chemical action so far considered is dependent upon the action of enzymes.

RT 41 (*a*) Review the process of protein synthesis.
(*b*) Review the structural and functional classification of proteins and enzymes.
(*c*) Review the action of enzymes and factors that affect their functioning.

8.4.3 Obtaining raw materials

Although both autotrophs and heterotrophs use organic molecules as their energy source, autotrophs require only inorganic elements as their raw materials for organic synthesis. The heterotrophs must obtain organic raw materials from other organisms.

RT 42 Review the ways in which autotrophs obtain their raw materials.

Because of their larger molecular size, organic macromolecules cannot be obtained by direct passage through the body surface. These organic raw materials must be degraded to their basic components for absorption and later use in the heterotroph's body.

Heterotrophs secrete chemicals onto their food which is degraded into small molecules that can pass through the organism's cell boundaries. This is achieved by extracellular chemical action **(digestion)**. There is a great variety of ways that heterotrophs use to combine digesting enzymes with organic food substances. Figure 90 shows the basic processes involved.

A variety of feeding mechanisms and lifestyles are employed to bring food into contact with the digesting enzymes. Figure 91 shows the different types of feeding mechanisms which are dependent on the type of food and its source. Feeding relationships are considered further in section 11.

90 Extracellular chemical action

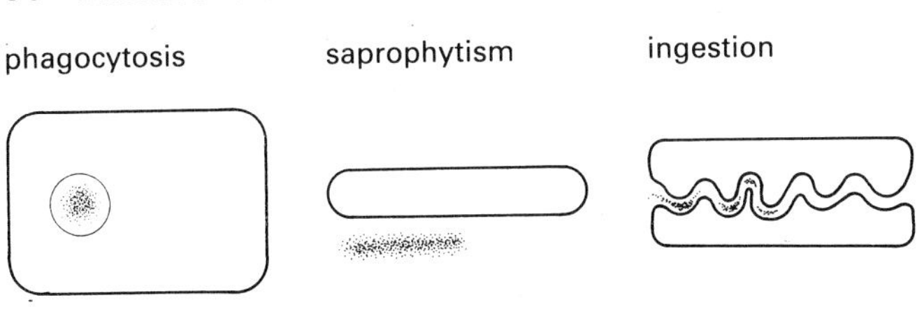

91 Types of heterotrophic nutrition

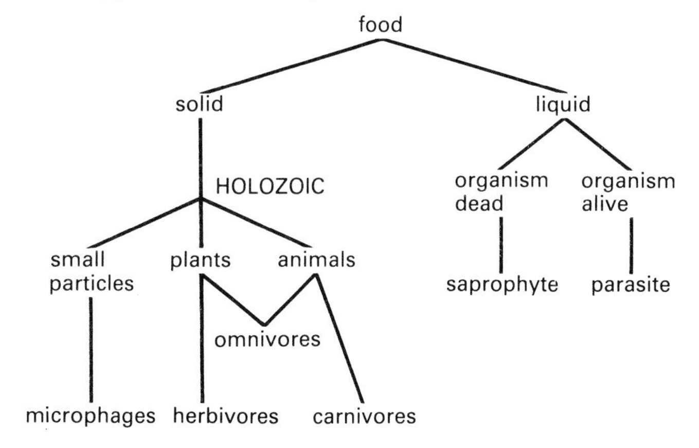

RT 43 (*a*) Review the chemical action of digestive enzymes on food material.
(*b*) Outline the differences in detail of chemical action seen in different types of heterotrophs.

Once the products of digestive action have crossed the boundary layers and the basic units are inside the heterotrophs, similar synthetic reactions occur, as detailed in section 8.4.2, and further degradations, as detailed in section 8.3.

8.5 Molecular movement

For efficient chemical action, the right chemicals need to occur in the right concentration at the right time and in the right place. This section looks at the methods of movement of chemical substances in and out of cells and around organisms.

8.5.1 Diffusion

One of the basic principles of movement of molecules in organisms is that all atoms and molecules are in constant random thermal motion. As a result, if substances have different concentrations in different places (that is a **concentration gradient** exists) molecules will tend to move in a definite direction until there is an even (or random) distribution. Such **diffusion** is fundamental for living things. At the cellular level, it is the main method of movement into and out of cells. Much of the organisation of organisms is concerned with the establishment and maintenance of such concentration gradients between the exterior and interior of cells.

RT 44 Review the processes of diffusion, osmosis and facilitated diffusion.

8.5.2 Active transport

Certain substances are needed in cells at higher or lower concentrations than occur in the surroundings of the cell. To keep such substances in or out, the cell should be impermeable to their passage. Molecular movement, other than diffusion is, therefore, needed to move these molecules into or out of the cell if their concentrations are not at the correct level.

RT 45 Review the details of active transport.

Proper membrane function is fundamental to life as activities of organisms are dependent on the processes occurring within their cells. Chemical action within cells is determined by cell membrane properties. Diffusion and active transport are the main ways molecules get in and out of cells.

RT 46 Describe how diffusion and active transport contribute to the functioning of the structures listed: lung/gill, root, leaves, kidney, capillaries, nerves, alimentary canal.

8.5.3 Mass flow

As organisms become more complex, size and activity of the organisms tends to increase. Because of the limitations imposed by the **surface area to volume ratio,** many organisms develop structures to increase their surface area without substantially increasing their volume. In larger or more active organisms, diffusion and active transport are not sufficient to supply the needs of all cells. Such organisms have developed a variety of methods to promote the **mass flow** of substances through or around their bodies.

RT 47 *(a)* Outline the reasons for the need of mass flow systems.
(b) Review the variety of mass flow systems found in living organisms and describe how they operate.

Molecules may also move in another way, by chemical action that causes certain molecules to slide past each other. These individual movements are relatively small and localised. This movement of actin and myosin is covered in section 8.6.3.

92 Fundamentals of response in living things

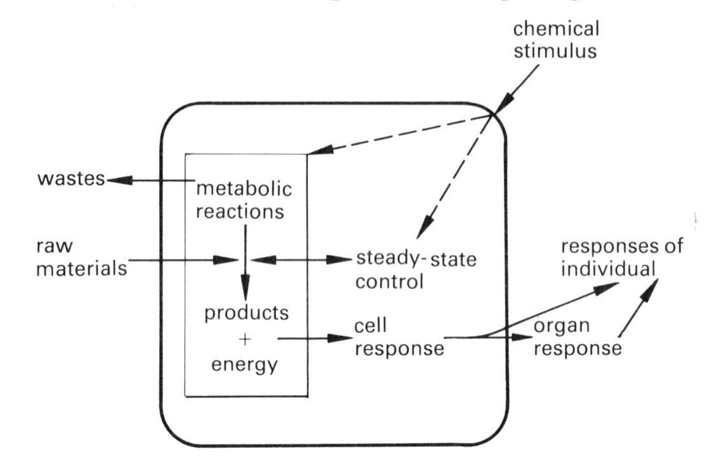

8.6 Chemical action initiating activities

Sections 8.1–8.5 have emphasised the chemical basis of life. Accordingly, all the activities of animals and plants can be viewed, in one way, as a change in the metabolic reactions occurring in cells. So far, metabolic reactions involved in the liberation of matter and energy have been considered. This section looks at a variety of chemicals and the activities they produce or initiate. The **control** and **integration** of these activities will be dealt with in sections 9 and 11.

Metabolic reactions occur within cells of the organisms. To initiate activities, stimulating substances must affect the supply of metabolites and/or affect the cell in some way to alter the reactions of metabolism (see figure 92).

The activity or **response** initiated does not occur in isolation or without some controlling influence. **Steady-state control** is a vital component in the response of living things (see section 9).

Receptors, whether single cells, complex organs or just parts of unicellular organisms, are subject to a variety of **stimuli**, not all chemical in nature. The function of these **receptive areas** is to convert the stimulus into some chemical change within the body of the organism. This change may then be communicated to other cells where a change in their metabolic processes results in a response by the organism as a whole to the initial stimulus (see figure 93).

8.6.1 Receptors

Invariably, the initial stimulus affects the functioning of the receptor cell membrane. Altering

93 Receptors and responses

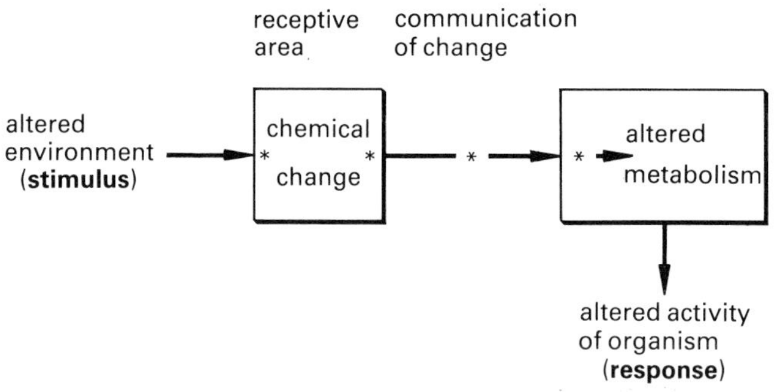

membrane functioning is the basic process involved in the reception and transformation of stimuli.

RT 48 Review the chemical changes that occur in receptive areas as a result of stimulation in animals and plants.

In many organisms, relatively simple receptor cells become associated to form **sense organs.** Other cells are also involved to protect the sensory cells and alter their sensitivity and general functioning. Despite the increased complexity of sense organs, the individual sensory cells function in the same way as the simple isolated receptor cells.

RT 49 (*a*) Review the classification of receptor cells according to the type of stimulus to which they respond and the structure of the receptor.
(*b*) Review the initiation of generator potential by stimulation of the receptor cell.
(*c*) Review the structure of sense organs sensitive to a variety of types of stimuli, and outline how the structures ancillary to the receptor cells protect and activate the receptors in named examples.

8.6.2 Nervous communication

The altered membrane functioning of receptors caused by the reception of a stimulus is communicated to the cells that cause a response **(effectors)** by nerve cells, here again the membranes play a key role.

RT 50 (*a*) Review the way generator potentials in receptor cells alter chemical activity of nerve cells and initiate action potentials.

94 The two types of cellular response

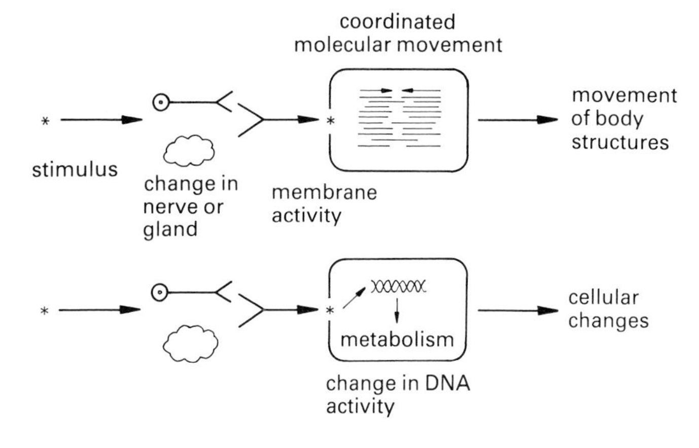

(*b*) Review the changes in nerve cell membrane functioning and chemical activities that cause the communication of membrane changes to *other* cells (other neurons or effector cells).

8.6.3 Effectors

The information concerning the reception of stimuli is communicated to cells that cause a response of the organism as a whole. The response is initiated by the release of chemicals which affect the membrane and then the metabolism of the cell itself. These chemicals, whether **nervous transmitter substances** or **hormones** bring about two broad types of response (see figure 94).

Note: hormone-producing organs can act as either communicator or effector depending upon the target of the hormones and the activity to be initiated in the organism.

RT 51 (*a*) Review the way hormones affect membranes to initiate alteration of DNA activity (and, therefore, metabolism) in cells.
(*b*) Review the details of chemical action in contractile tissue that bring about a response.
(*c*) Review the structure and functioning of a variety of effector organs in a range of animals.

When talking about nerves and muscles, we naturally think only of response in animals. It is important to remember that similar chemical action occurs in reception of stimuli, communication and response in plants.

Section 9 Equilibria

9.1 Introduction

Change is a basic process in Nature. In the face of change, living things maintain an active balance between themselves and their environment. The maintenance of this balance or **equilibrium** has to be an active or **dynamic** process to counteract the forces of change that are constantly operating.

Commonly the number of individuals in a population remains relatively constant. The structure and behaviour of offspring are very similar to their parents and the composition of internal conditions or mature ecosystems can remain constant for long periods. Such equilibria are maintained by **self-regulating** processes where the living system reacts to a disrupting influence in a way that tends to counteract the influence.

Living things have inputs and outputs of energy and matter. Different mechanisms exist to control the different forms of energy and matter. The mechanisms are self-regulating because the controlling structures are activated by **feedback** of information from **sensitive** parts of the mechanism itself that respond to changes in the inputs and outputs of the system. Figure 95 summarises such a mechanism.

95 General features of an equilibrium-maintaining mechanism

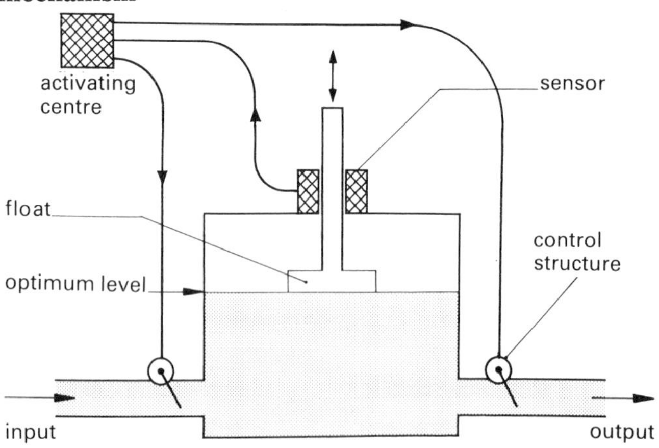

Notice the system is not closed, it is an open or dynamic system so that the optimum level or **set-point** is maintained in balance or equilibrium with the constant flowing in and out of matter and energy.

If the level of the system changes because of the forces that alter the balance of inputs and outputs, corrective measures are triggered to retain the equilibrium and restore the optimum.

An important feature of such mechanisms is that the levels are never constant, small fluctuations above and below the set-point occur. There is always a slight delay between the change from the optimum and the activation of the corrective measure. The actual level will **oscillate** about the set-point (see figure 96). The more efficient the self-regulating mechanism, the smaller the oscillations.

96 Change in level of a self-regulating mechanism

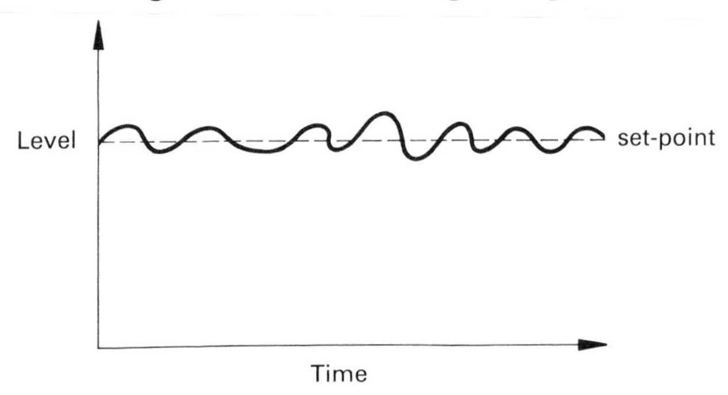

Also, most self-regulating mechanisms are capable of being adjusted to different set-points (see figure 97).

The actual biological details of such mechanisms will depend upon the feature being maintained at an optimum, such as concentration of body fluids or numbers and types of organisms in a group. You will be familiar with the term **homeostasis,** the maintenance of optimum **internal** conditions which is the fundamental process in living things. The

97 **Adjustment of the set-point**

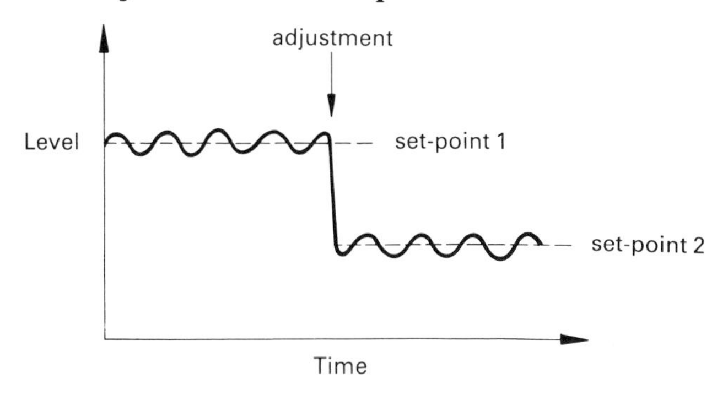

overall theme of this section, however, is **equilibrium**, the maintenance of a state of balance with the environment. Homeostasis is just one example of this broader principle.

9.2 Molecular equilibrium

RT 52 (*a*) Review enzyme stimulation and inhibition of enzyme action.
(*b*) Review alteration in enzyme activity by gene regulation.

Many biochemical reactions are controlled by negative feedback where an increasing product concentration of a reaction acts as an inhibitor on the enzyme catalysing the reaction. Similarly, the positive feedback from increasing substrate concentration stimulates the enzyme. A balance of the opposing forces maintains a constant product concentration in equilibrium with the needs of the cells. Feedback can affect either the activity of the enzyme itself or its concentration in the cell via the genes responsible for its synthesis.

9.3 Constancy of internal conditions

For life processes to proceed efficiently, the cytoplasm of cells must be maintained at optimum physical, chemical and biological conditions for chemical action. For example, constant temperature, pH, osmotic pressure, substrate and product concentrations, toxic and waste substances must be held at constant concentrations. The organism must also be free of disrupting influences such as foreign particles or organisms.

9.3.1 Physical and chemical conditions

In unicellular organisms, the physical and chemical conditions of the cytoplasm are maintained primarily by the action of the surrounding membranes and wall (if present). Some specialised homeostatic mechanisms do exist in such organisms.

RT 53 Review the action of homeostatic structures found in unicellular organisms.

In the majority of multicellular organisms there is a body fluid which is isolated from the external environment. The body fluid acts as a buffer between the cell and the external environment. There is an interaction between the cell contents and the surrounding body fluid (see figure 98).

98 **Cellular environments**

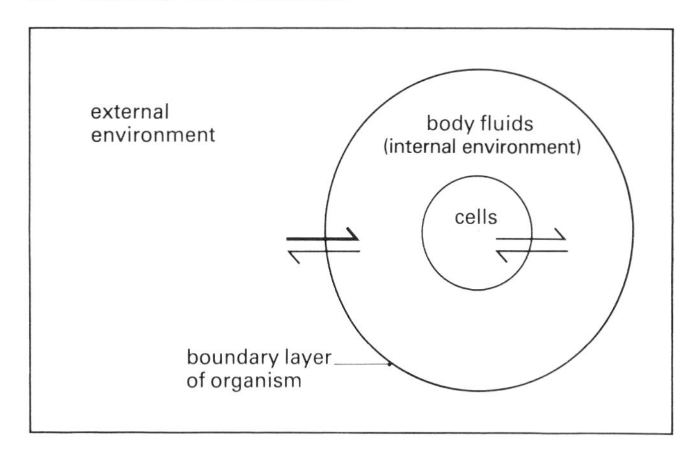

Maintenance of optima within such cells is largely achieved by controlling the body fluid composition. Figure 99 shows the systems involved in body fluid regulation.

99 **Equilibria systems in multicellular organisms**

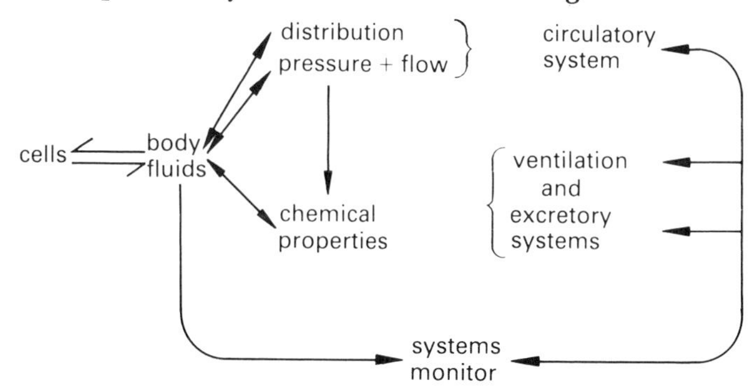

RT 54 For each of the parameters given below, review the way an optimum level is maintained in mammals (unless otherwise stated) using figure 100 as a model for your review.
Blood sugar, osmotic pressure (and fish), temperature (and reptiles), ions, respiratory gases, blood pressure and volume, hormones, ammonia (and insects)

100 Generalised homeostatic flow diagram

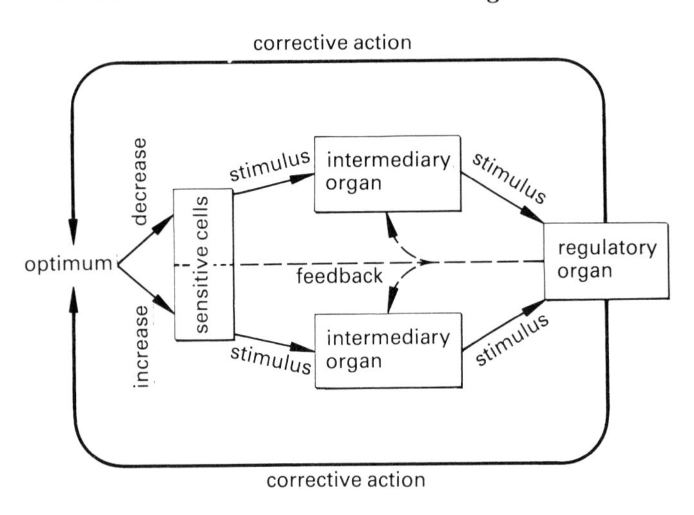

It is important to remember that these homeostatic mechanisms can only operate efficiently over a particular range of disrupting forces (see figure 101).

Different organisms and different parameters have different tolerance ranges outside which the organism is out of equilibrium with its surroundings. As long as the disrupting influence is not too

101 Range of efficiency of homeostatic mechanisms

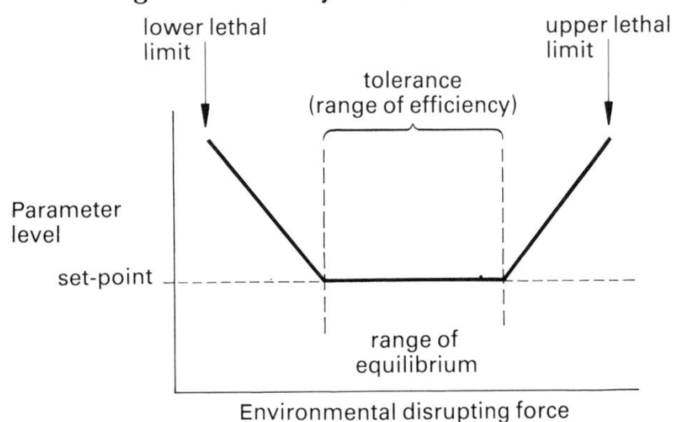

extreme (that is outside the lethal limits) or does not force the organism to be out of equilibrium for too long, the organism's mechanisms can recover steady states. If, however, the disrupting influence persists, the organism must either avoid the disrupting influence from the environment or its mechanisms must adapt to new set-points. If neither of these things occur, the organism will die from failure of its regulating systems.

RT 55 Review the tolerances of organisms to external temperature and osmotic pressure fluctuations and their responses to the fluctuations.

Internal equilibria result from the balancing effects of efficiently functioning tissue and organ systems within the individual. The maintenance of these optimum internal conditions can only be achieved if the external environment does not impose severe stresses on these systems. The structure and behaviour of organisms often arise as a response to the environment so that the amount of stress to which they are exposed is reduced. These characteristics help place the organism in as favourable an environment as possible, one in which the internal steady-state mechanisms can function efficiently.

RT 56 Review the structural and behavioural features of a variety of organisms that help the homeostatic mechanisms maintain steady states.

9.3.2 Biological integrity

An important concept in the maintenance of internal equilibrium is that the internal environment of an organism should be composed of cells and substances produced by the organism itself. Invasion by foreign particles (living or non-living) poses a threat to the normal workings of the organism. Many animals contain wandering phagocytic cells that remove foreign particles. Vertebrates have also developed a defensive or **immune system** to protect their internal environments from the disrupting presence of substances not formed by their own bodies.

RT 57 Review the mechanisms that exist for organisms to maintain the biological integrity of their internal environments.

9.4 Offspring stability

Reproduction is one of the characteristic features of living things. The basic process involves the formation of the genetic complement or **genome** of the offspring which usually involves some sort of cell division. Reproductive methods which allow variation in the offsprings' genomes are important because the offspring are thereby able to exploit the opportunities presented by a changing environment and, hence, survive. Section 10 considers this point further.

For the majority of time, most environments remain relatively stable. It is disadvantageous, therefore, for the offspring to be too different from their parents. The characteristics of the parents are successful in the prevailing environment and those successful characteristics produced largely by the genome should be passed on unchanged to the new generation. This **offspring stability** is obviously important for the survival of the individual.

For stability of characteristics, genes should be copied and passed on to offspring without being changed in any way. Both the method of gene copying (mitosis and meiosis) and the way these copied genes are passed on (inbreeding and outbreeding) affect how similar the offspring will be to their parents.

RT 58 For each point below, emphasise how genetic and offspring stability in the individual is maintained.
(*a*) Review the molecular mechanism of gene copying.
(*b*) Relate gene copying to cell division involving mitosis and meiosis.
(*c*) Review mechanisms of transfer of gene complements from parents to offspring.

Sections 9.1–9.4 have been mainly concerned with the stability of individuals. Sections 9.5 and 9.6 considers the stability of groups of organisms.

9.5 Population stability

Despite individual death, birth and migration, the population persists as a very stable, natural grouping. This section looks at some of the processes involved in the maintenance of this equilibrium between populations and their environment.

9.5.1 Population characteristics

Even in the face of possible individual variation in offspring, the characteristics of a population as a whole may remain constant over longer periods of time. Humans, for instance, have had many constant characteristics for over 40 000 years despite a reproductive method favouring variation in genome composition. This stability of inheritance shows that there must be some degree of genetic equilibrium. The genes of the individuals must be in balance with the environment. This stability is due to the way genes are passed on from generation to generation and the constraining influence of the environment on the genes inherited.

The passing on of characteristics from parents to offspring is predictable even when a particular gene can have more than one expression (its **allelomorphic forms**). For any two parents, definite ratios of offspring types occur. For the breeding section of a population, genes are available in definite frequencies for the production of the next generation.

These patterns of inheritance remain stable and constant as long as certain disrupting forces do not operate. Even if some extreme variants do arise in a population, the selective forces of the environment will remove any forms that are outside a relatively narrow environmental tolerance range.

RT 59 (*a*) Explain the occurrence of predictable ratios of offspring types produced by two parents.
(*b*) Explain the constancy of gene frequencies within populations.
(*c*) Outline the processes that may occur in (*a*) and (*b*) to disrupt their stability.

RT 60 Review the way the environment can act to maintain stable population characteristics.

9.5.2 Population size

A population is a self-regulating system, and another feature of the system that remains remarkably stable is the number of individuals or population size **(abundance)**.

Like the temperature or osmotic pressure of an organism, population size is maintained at an optimum level and in equilibrium with the environment by processes that tend to oppose changes in population size.

Maintaining a balanced population can be as important for the group as is maintenance of a balanced temperature for the individual. This steady-state control of populations is not termed homeostasis as the factor under control is not an internal bodily condition. Although stable, population size like any other self-regulated parameter, shows oscillations about the optimum level. In certain instances of climate or predation, the oscillations can be quite large, but an underlying balance and equilibrium is still being maintained.

The delicate and important nature of the balance can easily be seen when an event can upset a balanced population and cause a spectacular population explosion which, in turn, can affect the environment. The release of rabbits in Australia is one such example.

Figure 102 shows the two main factors involved in the balancing of population size.

102 Opposing forces in the balance of population size

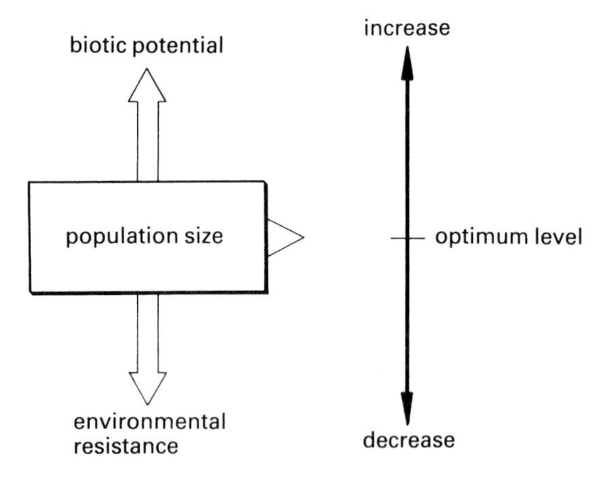

In the absence of any restraining influences, living things would show an exponential increase in numbers. It is extremely unlikely that such a situation would occur for long in natural environments. Reproductive capacity usually comes to be held in check by forces imposed from the biological and physical environments.

RT 61 Review the population dynamics of a group of organisms colonising a new habitat.

The restraining **environmental resistance** can be divided into two major components, **density-dependent** and **density-independent factors.** Density-independent factors are mainly due to effects of the physical environment which reduce population size by amounts related only to the severity of the prevailing conditions. The density-dependent factors (mainly effects of the biological environment) reduce population size by amounts that depend on the size of the population: the smaller the population, the smaller the effect and consequent reduction in numbers. These influences represent the self-regulating forces that maintain the steady state of population size.

The size of a population, therefore, results from the point of equilibrium reached between opposing forces. If any of the forces change, a new equilibrium will be reached and a smaller or larger population will result. As these forces tend to compensate for changes in one another, the population stays relatively stable (see figure 103).

103 Compensating effects in population regulation

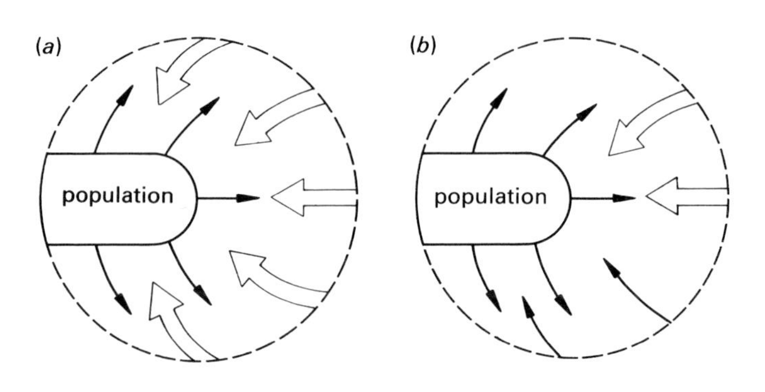

RT 62 Review how factors operating *within* a population **(intraspecific)** maintain population size.

Intimately linked with forces acting from within the population are forces originating from contact with other populations **(interspecific).**

For a population to be in balance, all other related organisms (predators, prey, competitors and parasites) must also be in equilibrium.

RT 63 Review regulation of population size by interspecific influences.

9.6 Ecosystem composition

Living organisms and the physical environment interact through the flow of energy and materials. When natural systems reach a sufficient degree of organisation we call them **ecosystems.** Ecosystems are comprised of a number of components (climate, soil, vegetation and animal populations) integrated and, to a greater or lesser extent, self-contained or **closed.** Establishing an ecosystem from bare ground involves **colonisation,** followed by a series of changes **(succession)** until a stable community, in equilibrium under the prevailing physical conditions, is established. This is the **climax community.**

RT 64 (*a*) Review the climax communities that develop in different physical environments.
(*b*) Review the forces acting within ecosystems to maintain the climax community.

Ecosystem stability can be maintained in the face of fluctuations in the regulating forces, but the balance is quite fragile especially if changes occur over relatively short periods of time. A major disrupting influence for ecosystem stability is the activities of human populations.

RT 65 Outline the ways human influence can affect ecosystem climax and stability.

Section 10 Adaptation

10.1 Introduction

Previous sections have considered the way self-regulation of living systems tend to control and maintain the system at optimum conditions in the face of fluctuations in the internal or external environment.

An organism's structural, chemical, behavioural and developmental characteristics arise largely from its genetic information under steady-state control within a particular environment. If the organism's characteristics function effectively and are well-fitted to the environment, the organism will be successful (survive and reproduce). Offspring show different degrees of variation in these beneficial characteristics depending on the method of reproduction and on any chance mutations that take place. If the environment is constant, the characteristics in a population will remain stable. If the environment changes so the individual's steady-state control mechanisms are not capable of counteracting the disrupting influences, two things may happen.

Firstly, an individual's characteristics or control mechanisms may **adjust** to operate with a new norm or set-point more favourable to survival in the new environment. Secondly, variants in the offspring that would not have been successful in the previous environment may now be able to operate effectively, be selected for and become successful. Either way, the individuals of populations adapt to or are adapted for the prevailing environment. These adaptations cause populations exposed to different environments to become different from one another. Eventually, different species will result. It is this inheritance of variation and selection of adaptive features that is termed **evolution**. This section looks at the origin of new characteristics, how they may become adaptive and at examples of adaptive changes that have occurred in living systems.

10.2 The origins of new characteristics

The characteristics of all living things arise by interaction between the genetic constitution (the **genome**) of an organism and its environment.

During the development of an organism, a changed genome or environment may result in modifications to the characteristics of the organism. It is a matter of chance whether the new characteristics will prove beneficial in the environment. A changed genome, the major source of new characteristics, can result from **mutation** or **recombination**. The method of reproduction largely determines the amount of recombination of the parental genome that will occur and, hence, the survival potential of the offspring. Reproduction and inheritance are, therefore, of great adaptive significance (see figure 104).

104 Recombination and survival

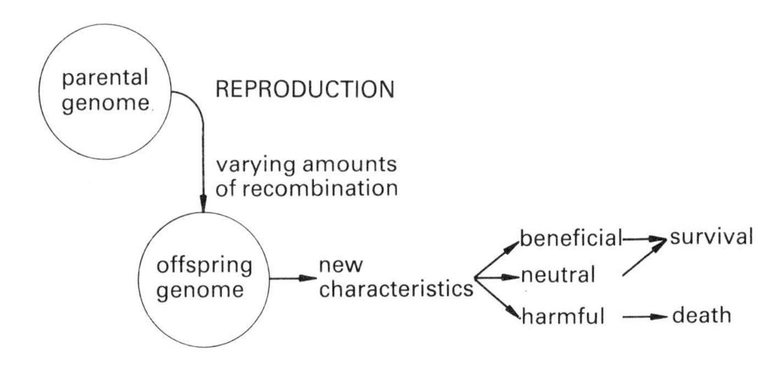

10.2.1 Reproduction of the genome

Asexual reproduction tends to reduce variation of the genome and, therefore, the offspring (this in itself can be of great survival value in an unchanging environment) whereas sexual reproduction by meiosis and fertilisation tends to promote variation.

RT 66 Review how meiosis can lead to variation in offspring genomes.

The extent of variation also depends on the way the gametes of the sexually reproducing parents are brought together.

RT 67 Review methods of fertilisation and explain how offspring genomes may be more or less variable in given examples.

Another source of genome modification is by **mutation**. Mutations can occur in body **(somatic)** cells with profound consequences for an individual organism. Of greater importance for an individual's offspring, and the population as a whole, are mutations in the genome of gamete-producing **(germ)** cells. It is these mutations that will be inherited by the offspring.

RT 68 Review the different types of gene and chromosome mutations.

10.2.2 Development of characteristics

Although all characteristics are ultimately controlled by the organism's genes, genes only represent the **potential** for a given characteristic. The **actual** expression of the genes is dependent on suitable environments.

RT 69 Review the role of the environment in the relationship between the genotype and the phenotype.

10.3 Variation to adaptation

We have seen how reproduction can give rise to variable genomes and variable characteristics. This section looks at how chance variations become adaptive features to enhance the survival of the individual and the population as a whole. The processes and consequences of **natural selection** are considered.

10.3.1 Natural selection

The environment affects the expression of characteristics but it also acts upon the characteristics themselves. There is a tendency for the environment to favour those characteristics that are best adapted to the environment. Such organisms tend to be more successful reproducers or survivors. Eventually, by posing too hostile an environment for less well-adapted organisms, the less advantageous characteristics disappear from the offspring generations (see figure 105).

It is this process of natural selection that turns variations into adaptive features. Disadvantageous variations tend to be replaced as a result of reproductive pressure.

RT 70 Explain how the genetic equilibrium of a population can be affected and the characteristics of the population change with time.

Progressive change in the characteristics of organisms in adaptive response to the environment is called **evolution**. The interplay of variation, inheritance and the environment leads to the origin of new species.

RT 71 Review the factors necessary for the formation of a new species.

RT 72 Review the development of the theories concerning the mechanism of evolution and the evidence used to support them.

Some organisms can also adapt to their environment by modification of their behaviour in light of past experience, in other words by **learning.**

Unless advantageous behaviour patterns become **innate** and under genetic control, there is little survival value for the offspring as the learning has to be acquired afresh. A relatively safe way for this learning to take place is by the offspring being **taught** by the parents or by the offspring learning

105 Survival of the fittest

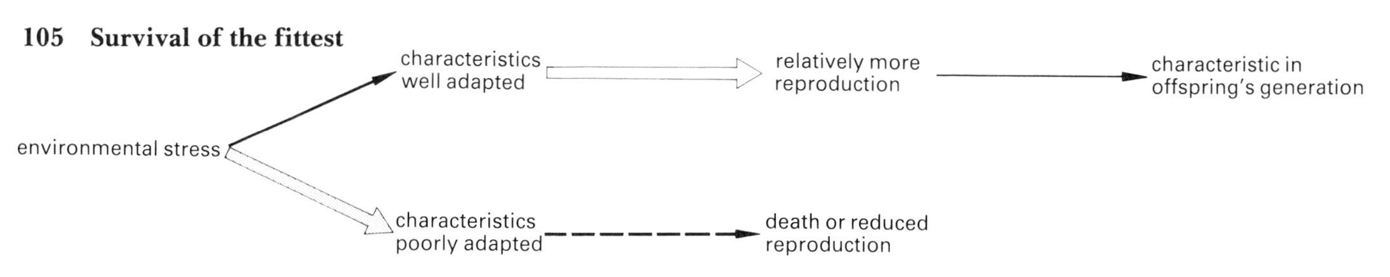

the behaviour itself while under the protection of the parents.

RT 73 Review the different types of learning seen in animals.

RT 74 Review the role of parental care in survival of offspring.

It is important to realise that there is no most highly evolved or perfectly adapted organism or group that evolutionary processes are 'aiming' for. Evolution is not geared toward the ultimate goal of producing the human species. Evolution is really random opportunism. Those species alive today are not, theoretically, the best or most desirable forms, but they represent the most practical possibilities available. Evolution produces adequate, workable alternatives. Some examples of this evolutionary opportunism are considered in the next section.

10.4 Some major adaptations of living things

The major adaptations of living things have occurred in the course of evolutionary sequences that are still continuing. Organisms within the same adaptations are either closely related (share a common ancestor) or have developed the same adaptive solution to similar environmental problems.

For example, ferns and vertebrates are obviously not closely related but they both have the adaptation of motile male gametes.

RT 75 Review convergent and divergent evolution and explain the difference between homologous and analogous structures.

Natural classification attempts to express the evolutionary relationships of living things **(phylogeny)** based upon the **homology** of organs. This takes into account the evolutionary history and origins of living things. Natural classification can, therefore, be seen as an evolutionary map charting the rise and fall of adaptive characteristics (see figure 106).

RT 76 Consider the evolutionary sequences outlined in figure 106. Under the headings listed below, comment on the major adaptations shown in each sequence.
Water conservation, exchange surfaces, nervous systems, reproductive methods, musculoskeletal systems, transport systems, life cycles, nutrition

106 Evolutionary trees for (*a*) the Plant Kingdom and (*b*) the Vertebrata (the width of the graph for each group indicates the changing abundance with time)

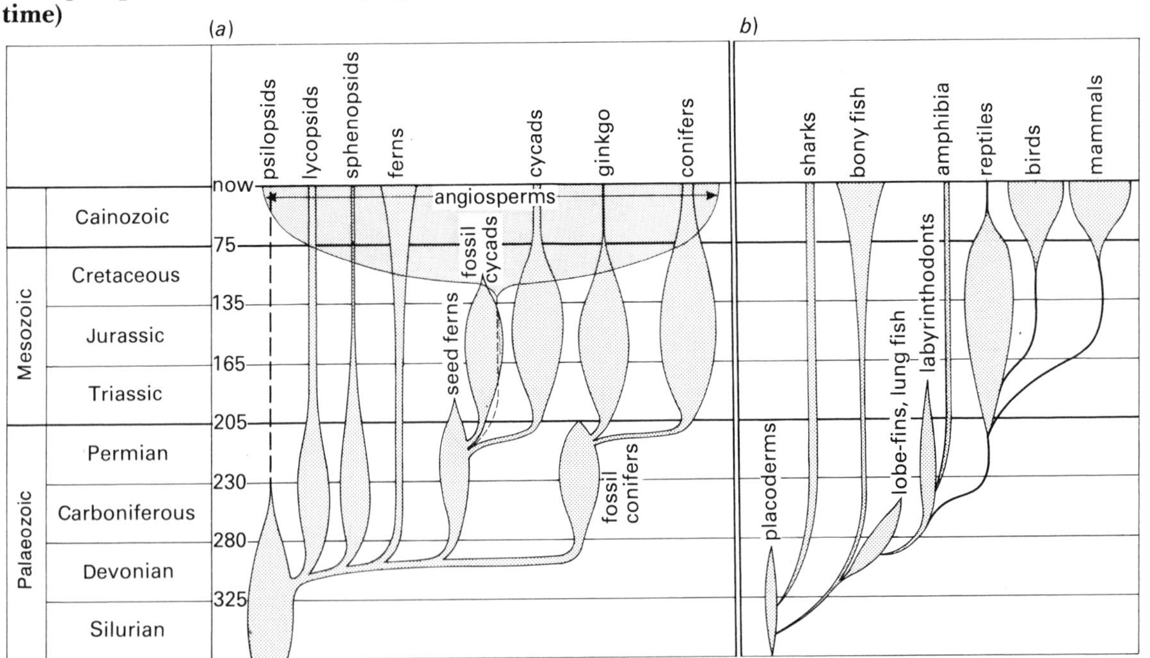

Section 11 Interaction

11.1 Introduction

This last section considers the most fundamental of themes in biology. All facets of the major themes so far considered are dependent on **interaction** between organisms or parts of organisms. Interactions of one form or another play a part in every event that occurs within living systems.

Perhaps the most basic of all is the interaction between the DNA and cytoplasm that organises the formation of a mature individual. Once an organism has grown and developed, it has to function within the constraints of the environment. Interaction of structures within the body ensure optimum internal conditions for life processes.

Interaction of the organism with its physical and biological surroundings is required to avoid harmful situations, to ensure supplies of energy and raw materials and for the production of offspring.

107 Major interactions

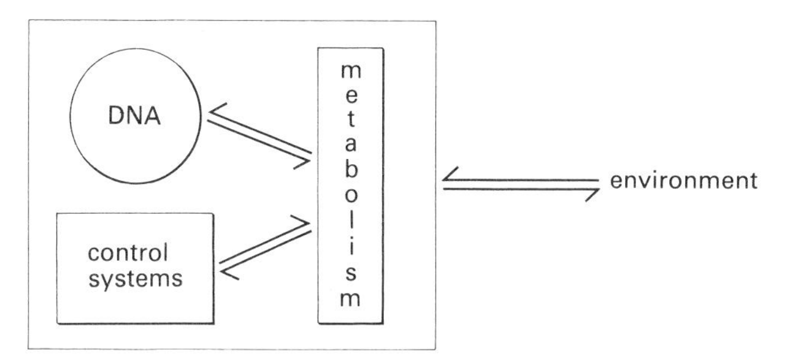

11.2 Interaction producing life

DNA is the template from which all living structures and processes arise, grow and develop. It is DNA's initiating and controlling effects that will be looked at in this section.

RT 77 Review the changes in form that plant and animal embryos go through during early development.

As the embryo grows, tissues and organs develop. Interaction between the embryo and its surroundings at a certain point in development cause the embryo to emerge as a new individual.

RT 78 Review the patterns of development of a variety of animals and plants, indicating which tissues and organs develop and what factors bring about emergence of the embryo (for example, xylem, phloem, bone, and so on) seed generation types and factors that promote birth, and so on.

11.2.1 Interaction in embryos

Life usually starts as a single cell containing DNA in the nucleus, cytoplasm and some organelles. The DNA initiates the formation of more cells by the processes of **replication** and **protein synthesis** until a ball of cells is formed.

Cellular interactions and movements (**gastrulation** in animals) turns this ball of cells into an early embryo with a structure characteristic of the particular type of organism. At the same time, the cytoplasm of the cells is interacting with the DNA to affect the type of protein synthesis taking place. This **gene regulation** causes the **differentiation** of different cell types in the embryo.

11.2.2 Maturation

Once the embryo has emerged into the world it grows and undergoes a series of changes in both structure and function until an adult form is reached. These 'life changes' are characteristic of different plant or animal groups and are related to the organism's environment.

RT 79 Review the variety of life cycles with emphasis on changes in growth, structure and function of the individual.

11.3 Interactions within the individual

Control and coordination of tissue/organ systems to maintain optima, initiate activities at certain stages in development, movement and locomotion.

RT 80 Review biochemical pathways, internal drives producing behaviour (programmed innate patterns too).

11.4 Interactions between individuals

RT 81 Review feeding, reproduction, social organisation, animal associations, communication, competition and so on. Behaviour in general.

11.5 Interactions between living things and the environment

RT 82 Review matter and energy movements, water relations, balance of oxygen and carbon dioxide. Environmental control of photosynthesis, biorhythms/cycles.

Section 12 Answers to self-assessment questions

1 The term selection pressure may be defined as any factor of the physical or biotic environment which leads to the survival of some organisms, better than others.

2 The biological meaning of the term 'fitness' is the extent to which the genotype is passed from generation to generation.

3 (*a*), (*d*), and (*e*) are biotic; (*b*), (*c*) and (*f*) are abiotic.

4 The term altruism may be defined as that characteristic of an organism which causes it to behave in such a way as to increase another organism's welfare at the expense of its own.

5 Any action which is carried out for the benefit of another can be 'costed' in terms of energy, time and so on. Such expenditure can be seen to be a diversion of resources away from the organism's assumed goal (that is, survival to reproduce maximally).

6 Homeostasis may be defined as the maintenance of equilibrium between the organism and its environment (internal and external).

7 Natural selection may be defined as those processes which occur in the environment which result in the survival of those organisms best adapted to the pressure of change, and the death of those least able to adapt to the change.

8 Some factors which contribute to evolutionary change are natural selection, mutation pressure, genetic drift, gene flow.

9 The main observation of Bernard was that the internal environment of the mammal was of a fixed constitution, and that its maintenance was a dynamic process.

10 The main factor which causes disturbances in the composition of extracellular fluid is the metabolic activity of the cells which it surrounds.

11 (*a*) Telphusa is the regulator and *Maia* is the conformer. (*b*) The regulator can withstand the wider range. (*Maia,* the sea crab, is not adapted to cope with low external concentrations.)

12 The advantage of an animal which can maintain internal homeostasis over one which cannot is that it is relatively more independent of its environment because it can make adjustments in response to external changes before their effects can have a damaging effect.

13 Dynamic equilibrium is the maintenance of the constancy of composition by the continual addition and removal of the constitutents. Static equilibrium is the maintenance of constancy of composition in a closed system. Water in a bottle or the elements in a crystal would be in a state of static equilibrium.

14 The three main components of a homeostatic mechanism are input, comparison and output.

15 Two (see figure 108).

108 Answer for SAQ 15

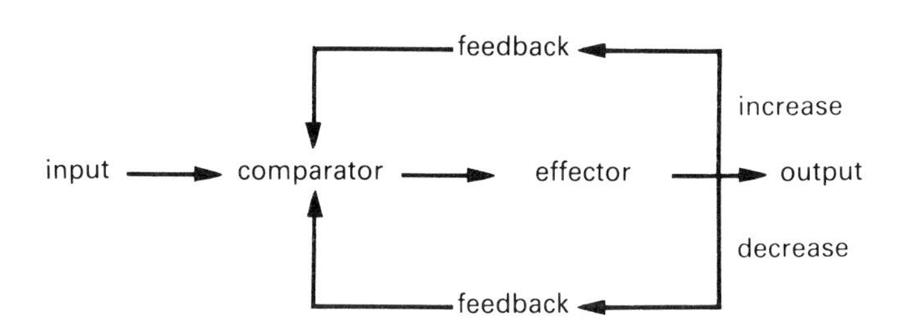

16 See figure 109.

17 See figure 110.

109 Answer for SAQ 16

volume of extracellular fluid

comparator: fluid volume receptors in the heart and kidney

thirst | ADH release | hormone release from kidneys and adrenal glands

drinking | decreased water excretion via kidney | sodium excretion via kidney

increase | increase | decrease

negative feedback loop

volume of extracellular fluid

110 Answer for SAQ 17

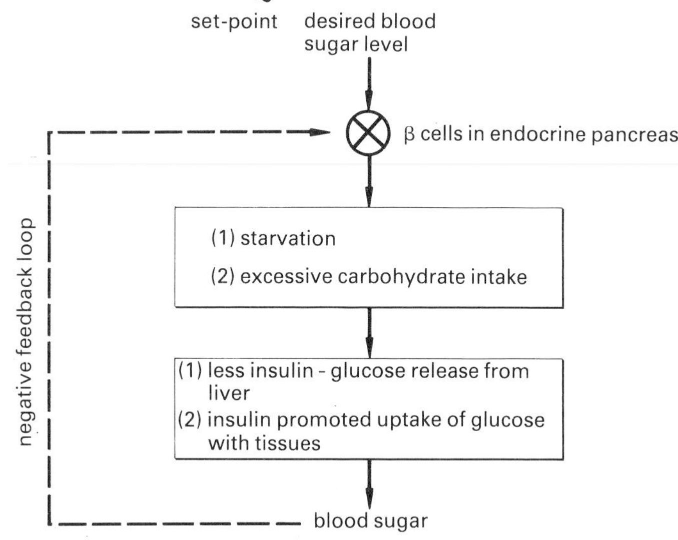

18 The oestrogen/luteinising hormone system provides an example of positive feedback in the following way. FSH stimulates the ovary to secrete oestrogen. The circulating level of oestrogen in the blood builds up because oestrogen stimulates release of LH which, in turn, stimulates release of more oestrogen until the surge of LH reaches a critical point at which the Graafian follicle is caused to rupture, and the ovum is released. The feedback is therefore positive, because the two hormones are enhancing the effect of each other.

19 The upstroke (rising phase) of the neuronal action potential provides an example of positive feedback in the following way. Change in the potential across the membrane on stimulation increases the permeability of the plasma membranes, and sodium ions flow in. This changes the membrane potential and sodium permeability is increased, so more sodium flows in, and so on. The feedback is positive, because the increase in permeability is enhanced by the change in membrane potential.

20 See figure 111.

111 Graph of heat production against environmental temperature for *Eulampis*

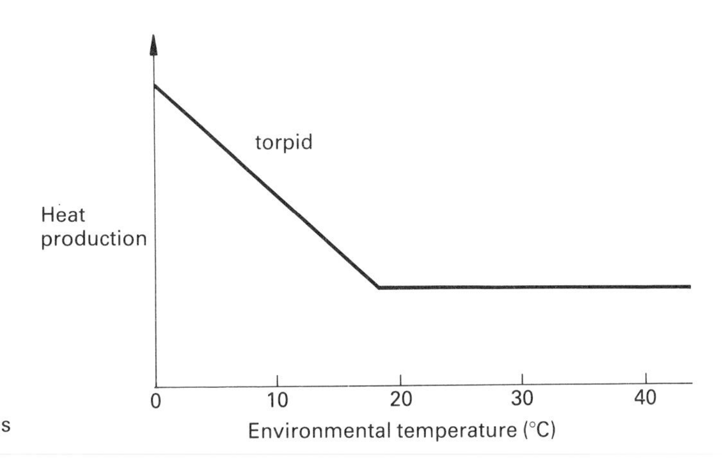

21 (*a*) Cuts down evaporation.
(*b*) Reduces evaporation losses.
(*c*) Minimises water loss.
(*d*) Can cool down at night a little and so afford to warm up during the day.
(*e*) Provides insulation from the Sun's rays.
(*f*) Can maximise use of infrequent watering points.

22 Heat is conducted from the incoming water to the outflowing water so that in the steady-state condition the outflowing water is prewarmed to within one degree of the incoming water.

23 The responses of humans to hypoxic conditions are hyperventilation and an increase in number of red blood cells.

24 Respiratory alkalosis results when the pH of body fluids is increased. The cause of this is an increase in the rate of removal of carbon dioxide from the blood due to hyperventilation.

25 See figure 112.

112 Answer for SAQ 25

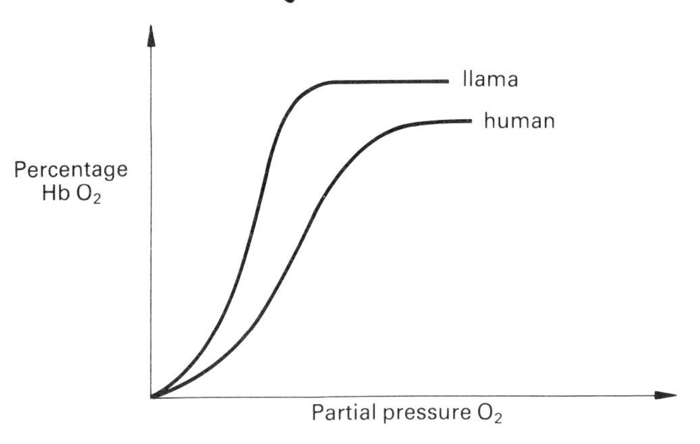

26 The respiratory and circulatory responses of the diving mammal are:
(*a*) increased oxygen storage by tissue myoglobin and increased blood volume;
(*b*) lactic acid metabolism for anaerobic respiration;
(*c*) decreased oxygen consumption from decreased metabolic rate during diving;
(*d*) decreased heart-beat rate;
(*e*) vasoconstriction (to maintain blood pressure) which redistributes blood to vital organs.

27 See figure 113.

113 Answer for SAQ 27

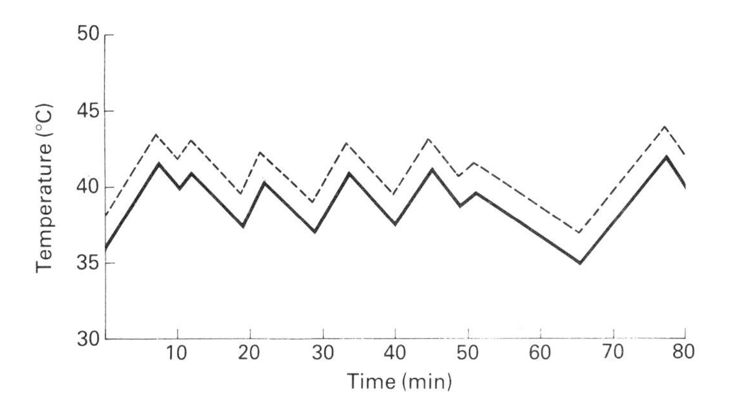

28 More examples of thermoregulatory huddling behaviour are shown by desert rodents, Antarctic penguins, newly hatched chicks in a nest.

29 The annual range of soil temperatures in a burrow of 1.5 m depth is approximately 58% of that at the surface.

30 (*a*) Uses heat from the Sun to raise its body temperature in the morning.
(*b*) Reduces conduction of heat from the hot ground to the animal.
(*c*) Reduces water loss by evaporation.
(*d*) Reduces excretory water loss.

31 The following can be offered as explanations for the phenomenon of migration.
(*a*) ecological – competition for niches, varying yearly predation
(*b*) historical – (i) moving land, mountains or seas,
(ii) 'disappeared' intermediate areas
(*c*) physiological – breeding areas may give specialist diets for rearing young
(*d*) environmental – climatic factors

32 See figure 114.

114 Answer for SAQ 32

Environment	*Adaptation*
Abiotic	blood of llamas
	blubber in whales
	water receptors in seals
Biotic	carnassial teeth
Deme	white flashes on antelopes' tails
Internal	loops of Henlé in kangaroo rat

33 The blood of llamas can be said to be a genetic adaptation, because llamas bred in zoos for many generations still retain the ability to take oxygen at low partial pressures into their blood and would be able to live at high altitude.

34 The homeostatic regulation of body temperature in a mammal provides an example of physiological versatility, because the constant temperature of the body is maintained in the face of changing environmental conditions by the homeostatic mechanisms for heat gain or loss (such as shivering and sweating).

35 One of the reasons that the panda is declining in numbers is that it is a selective feeder which is physically and behaviourally dependent on one major food species, bamboo. It cannot tolerate a varied diet. So in the wild, if its food species becomes limited or unavailable, the animal's survival is under immediate threat. Omnivorous humans, on the other hand, are highly adaptable where diet is concerned and will eat almost anything with any nutritional value if under sufficient pressure.

36 (*a*) Adaptive in conditions of complete darkness; bioluminescence is used for species recognition, sometimes as a lure to catch prey and sometimes to confuse predators.
(*b*) Pressure at great depth would crush a swim bladder; many deep sea fish have adapted to do without one.
(*c*) Temperature at depth is constant but low; fish have adapted accordingly.

37 Developmental flexibility is shown by humans in response to long-term exposure to high altitude because the number of red blood cells per unit volume increases. This returns to normal on return to sea level.

38 Primates are at an advantage as a result of their ability to learn from experience because on re-encountering an unfavourable situation, they are able to avoid it or, perhaps, even to manipulate it.

39 Adaptations for diving in the seal include:

physiological – high blood volume
– high haemoglobin content
– high tissue myoglobin
– use of lactic acid in anaerobic respiration
– decreased metabolic rate
– interrupted supply of air to prevent 'bends' and oxygen toxicity
– reduced heart-beat rate

structural – water receptors on face to keep nostrils closed
– blubber
– streamlined body and flippers
– countercurrent exchangers in flippers to conserve body heat

40 *Notonecta* and *Corixa* both occupy the freshwater habitat but, because their feeding activities and related behaviour are different, they are said to occupy different niches.

41 The earliest dated land animal fossil is not necessarily an example of the first type of land animal because the chances are that the type of organisms which first emerged on to land probably did so in small groups of relatively unspecialised forms. The fossil record consists mainly of large populations of specialised forms which flourished freely until they became extinct when conditions changed.

42 The Eocene Period was 36–58 million years ago.

43 *c d b a e*

44 *Merychippus*

45 About 10 Myr ago

46 Extinction

47 Palaeontologists suggest either climate change or activities of humans.

48 They have been reintroduced, beginning with the Spanish in the sixteenth century.

49 The Carboniferous Period is so called because it was the time of the laying down of the major coal seams and thus the storage of carbon-containing fossil fuels. It was characterised by huge tropical fern-like trees, the first-known mosses, seed-ferns and conifers, amphibians (dominant on land), a great variety of insects and tropical seas.

50 Structural changes associated with the succession of plant types include:
mechanisms which would have enabled them to take in water and minerals and allow for gaseous exchange from the water as they began to invade the land
independence of water for the purpose of reproduction
systems for support, such as woody stems

51 The succession of organisms that invaded the land is likely to have been:
plants (mosses to trees)
terrestrial invertebrates (arthropods and molluscs; insects in great variety)
terrestrial vertebrates – amphibia, reptiles, birds, mammals

52 Breathing gaseous oxygen and transport of gametes.

53 Development of an enclosed egg: a leathery but permeable shell protects the actual ovum from mechanical damage and desiccation.

54 See figure 115.

115 Answer for SAQ 54

Metatheria	*Eutheria*
Thylacinus	*Canis*
Dasyurus	*Felis*
Petaurus	*Glaucamys*
Phascolomys	*Marmota*
Myrmecabius	*Myrmecophaga*
Notaryctes	*Talpa*
Dasycercus	*Mus*

55 This similarity of animal types with different ancestors is an example of convergent evolution. Similar selection pressures may act upon genetically separate groups, bringing about independent, but similar adaptive evolutionary changes.

56 The orders which have fewer than 10 species are:
Dermoptera (2)
Sirenia (5)
Proboscidea (2)
Hyracoidea (5)
Tubulidentata (1)
Pholidota (7)

57 See figure 116.

116 Answer for SAQ 57

Order	*Adaptations for feeding*	*Adaptations for locomotion*
Artiodactyla	open rooted teeth; ridged molars; long gut; rumen	long, light legs; large shoulder and hindquarter muscles; unguli-grade hoof
Primate	generalised omnivorous dentition; binocular vision (good location)	opposable digits; prehensile tails (some); upright stance (man); binocular vision (distance judgement)
Carnivora	carnivorous dentition; snapping jaws; binocular vision (hunting); short gut	digitigrade; sprinting (cats); long-distance running (dogs)
Insectivora	acute senses (jaw); location; carnivorous dentition	
Chiroptera	blood-sucking: mouth-parts, anti-clotting saliva; insect-catching: senses; fruit-eating: dentition	forelimbs modified to wings
Rodentia	gnawing teeth	

58 See figure 117.

117 Pie chart representing floral species in Tristan da Cunha, South America and Africa

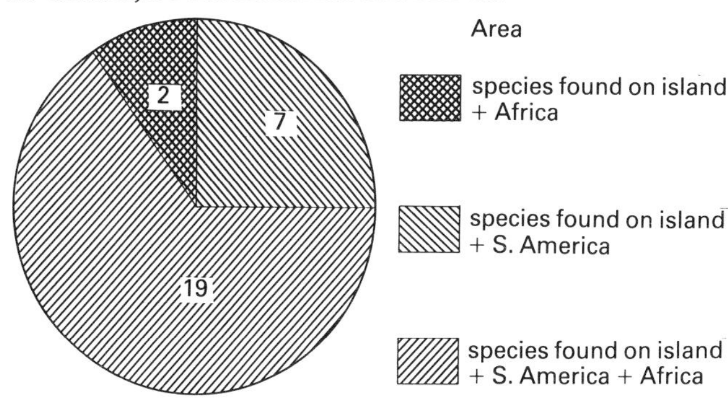

59 The Lamarckian explanation is as follows. The fish had eyes, but as they got cut off from light in an underground cave their eyesight became less functional, and this poor eyesight was passed on from generation to generation until they were completely blind. This process continued until their eyes were no longer present.

60 'Intraspecific' means within species.

61 Kettlewell's results show:
(*a*) that more dark moths than light moths were recovered in the industrial area (Birmingham);
(*b*) that more light moths than dark moths were recorded in the rural area (Dorset).

62 (*a*) Banded light shells are more difficult to see in the hedgerows.
(*b*) Dark, unbanded shells are more difficult to see in the woodland.

63 It is selectively advantageous to possess the sickle-cell genotype only in the heterozygous condition. People with HbA/HbS genotype are said to have sickle-cell trait, and this is associated with resistance to a particular form of malaria. They are, therefore, at a selective advantage in areas where malaria is endemic.

64 The lesser black-backed gull and the herring gull are different species, and generally do not interbreed. Starting with the herring gull and travelling west around the world, there is a range of variants of gull until, finally, right around the globe the gulls are so different from the original type that they are a distinctly different species (that is the lesser black-backed gull).

65 Reproductive isolation means the occurrence of mechanisms that prevent interbreeding.

66 The conclusion that the female *Drosophila* was unable to detect 'wrong' courtship patterns after her antennae had been removed can be criticised because the assumption is made that no other function is impaired by the loss of the antennae. It would be difficult to devise a proper control for this type of observation.

67 St Bernard's could interbreed as could the chihuahuas, but the two groups would be mechanically isolated from each other. Therefore, the two groups would probably remain distinct, possibly eventually becoming two distinct species.

68 Beak **1** is adapted to eating food **B**. Beak **2** is adapted to eating food **D**. Beak **3** is adapted to eating food **A**. Beak **4** is adapted to eating food **E**. Beak **5** is adopted to eating food **C**.

69 The absence of predators and competitors on the Galapagos Islands meant that the selection pressures were not so great for the ancestral bird that reached them. The geographical isolation of the individual islands allowed the establishment of new species.

70 Statement **1** refers to diagram **B**. Statement **2** refers to diagram **C**. Statement **3** refers to diagram **E**. Statement **4** refers to diagram **A**. Statement **5** refers to diagram **D**. Statement **6** refers to diagram **F**.

Index

Numbers in italics signify figure references.

acclimation 4
adaptation 18–20, *31*, 88–90
adaptive radiation 31, 32
adenosine triphosphate (ATP) 74, *74*, 76
aestivation 4
age of reptiles 32
alarm calls 3, 55
allomones 56
amensalism 59
Amoeba, encystment of 4
Anabaena 52
angler fish (*Photocorynus spiniceps*) 48
anoxia 13
antidiuretic hormone (ADH) 9
antipredator devices, plant 55
aphids 2
Archaeopteryx 22, 25
autotrophic bacteria 78, *78*
Azolla 52

bacteriophages 59
bats 32
behaviour 3
Bernard, Claude 5
biomes 70
biological integrity 84
biosphere 70–1, *71*
bombykol 48–9
bracken 55
bradycardia 13

Calliactis sp. 54
Calvin cycle *77*, 78
camel 12
camouflage (cryptic colouration) 55
cell 67
Cepaea nemoralis 39–40, *40*
chemicals 72–3
  functions *72*
  inorganic compounds 72, *72*
  organic compounds 72–3
Chirocephalus, egg dormancy 4
chlorophyll-less plants 59
chloroplasts 52, 76
circadian rhythm 3
climax community 87
clovers 55
coelom 68
coevolution 59, 60, *60*
colony, concept of 47
colouration 55
commensalism 54
communities 70
competition 31
conformers 6
constancy of internal conditions 83
copulins 53
Corixa 21
countercurrent heat exchange 12, *12*
courtship procedures 49
crickets 49
crustecydsone 50
cyanide 55
Cynognathus *32*

Darwin, Charles 1–2
  On the origin of species 39
  voyage on HMS Beagle 38, *38*
dehydrogenation 73–4
demographies 50
descent with modification 21
desert soil temperatures *16*
developmental flexibility 20
diapause 4
dinosaurs, extinction of 32
diploblastic animals 68
diving mammals 13–14, *13*
dodder 58
dolphin 12, *12*
Drosophila 49
dynamic equilibrium 7

Earth, age of 28, *28*
ecosystem 70, *70*, 87
effectors 81
Egyptian plover (*Pluvianus aegyptus*) 54
endoderm 68
energy liberation 73
environment 2
enzymes 78
Eohippus (*Hyracotherium*) 23–4
equilibria 82–7, *82*, *83*
eukaryotic cells, origin of 29
Eulampis jugularis *11*, 11
Eutheria 32, *33*, *34*
evolution 36–45, 89, 90
  anatomical evidence 36
  biogeographical evidence 37
  fossil evidence *see* fossils
  selective breeding evidence 37
  taxonomic evidence 36
evolutionary tree *90*
extracellular chemical action *79*
extracellular fluid 5
  homeostatic control of volume *8*

feedback 8–9
fever 14–15, *14*
fitness 1–2
follicle stimulating hormone (FSH) 9
fossils 21–30
  extinction 29–30; blanket (total, global) 31; coextinctions 31; local 30; taxonomic 30
  history of life revealed by *23–4*
  replacements 29
Fucus serratus 49

Galapagos finches 44–5, *44*, *45*
genome 88–9
geochronological aeons 28
Ginkgo leaf, fossilised *30*
giraffe 37–8, *37*
glycolysis 74
Graafian follicle 9
Gunnera 52

harvestmen 54
Heliconius butterflies 60–1, *60*, *61*
herbivory 55
hermit crab (*Dardanus*) 54
heterothermy 11, 12
heterotrophs 79, *79*
hibernation 4, 11
holly leaf miner (*Phytomyza ilicis*) 57
homeostasis 5, 7–8, 82, *84*
homeothermy 10–11, 15–16
  homeostatic maintenance *10*
honey-bee 3, 16
hormones 81
horse 23, *26*, *27*
hydrogen *75*
hyperventilation 12
hypothalamus 15
hypoxia 12

immune response 77
immune system 84
individual, concept of 47
insects 55
interactions 46–61, 91–2, *91*
  between individuals (intraspecific) 48, 92
  between populations 51–9, *51*
  between taxa 59–61
  in embryos 91
  male-female 48
  male-male 49
  within individuals 92
  within populations 46–7
internal environment, influences on/by *6*

kairomones 56
Kettlewell, H. B. D. 39
Krebs cycle *74*
Kronborgia amphipodicola *56*

labyrinthodont 31–2, *32*
Lamarck, Jean Baptiste 37-8
laurel 55
lemur 50
lichens 52
life, definition/characteristics 63–5, *63*
light-dependent reactions 77
living system 63
lizard, body temperature 15, *15*, 16
llama 13
luteinising hormone (LH) 9

macaques 50
Maia, blood concentrations *7*
mammals 32
  homologous forelimb structures *36*
  monophyletic ancestor 32
  polyphyletic ancestor 32
  representatives of most successful orders *35*
man, origin of 22
mastodont 31, *31*
maturation 91
Merychippus 23, 26
mesoderm 68
Mesohippus 23, 26
Mesozoic age 29–30
metabolism 72
Metatheria 32, *33*
migration 16, 17
milieu intérieur 5
mimicry 54
mineral salts 72, *72*
Miohippus 23
mistletoe 58
Mixotricha paradoxa 53, 53
molecular movement 79–80
mussel (*Mytilus edulis*) 54
mutation 89
mutualism 52
mycorrhiza 58–9, *58*

Nannipus 26
natural classification 90
natural selection 39–40, 89
  computer simulation 40
  dynamic theory 40
  kinematic theory 40
nervous communication 81
nervous transmitter substances 81
neutralism 51
niche 20–1
Nostoc 52
Notonecta 21

oestradiol 9
offspring stability 85
oral contraceptive pills 9
organ 68, 69
organelles 67
organic compounds, incorporation in energy-
  liberating reactions *75*
organisation 66-71
  cellular 66–7
  ecosystems 70
  group 69–71
  levels *66*, *67*
  multicellular level 67–8
    colonial level 67-8
    tissue level 68
organism, individual 67
organ systems 69
Orobanche 58
orthogenesis 26
oxidative phosphorylation 74
oxpecker (*Busephalus erythrorhynchus*) 54
oxygen
  adaptations to decreased availability 12–13
  dissociated curve (human) 13

palaeontology 21
palolo worm 48
Paramecium aurelia 53
parasitism 56–9
Passiflora spp. 60, *60*, 61
pea-crab (*Pinnotheres pisum*) 54
peppered moth (*Biston betularia*) *18*, 21, 39, *39*
Peripatus 23
pheromone 49, 49–50, 56
phosphoglyceraldehyde *77*, 78
photophosphorylation 74, 76-7, *76*
photosynthesis 78
physiological versatility 20
Piptoporus betulinus 58
plant(s) 3
  seed dormancy 4
plant kingdom, history of *30*
plant–phytophage coevolution 60, *60*
poikilotherms 16
polyploidy (abrupt speciation) 41–2
population 50, 69–70
  characteristics 85
  compensating effects in regulation 86
  size 85–7
  stability 85
Portuguese man-of-war (*Physalia*) 47–8, *47*
predation 54–6
Pronuba yuccasella 54
protocooperation 53–4
pyruvic acid dehydrogenation *74*

rearrangement reactions 74
receptors 80, 81
recombination 88, *88*
red fox (*Vulpes vulpes*) 53
redox reactions 73
regulators 6
relict orders 34
Remora remora 54
reproductive isolation mechanisms 42–3
reservoir, water level maintenance *7*
respiration 76
respiratory alkalosis 12
respiratory quotient 76
Rhesus monkey (*Macaca mulatta*), females 53
ring species of gulls *41*

Schistocephalus solidus 58, *58*
Schistosoma mansoni 57, *57*
sea anemone (*Anthopleura elegantissima*) 50
seal, heart-beat rate during dive *13*
sea slug (*Navanax*) 50
selection
  group vs. individual 3
  kin 3
  natural 1
semiochemical 56
sense organs 81
silk moth 48–9, *48*
Smilodon 31, *31*
societies 50
sociobiology 50
speciation 39, 40–2
  abrupt (polyploidy) 41–2
  allopatric 41
  parapatric 41
species 69
Spencer, Herbert 2
stromatolites, living in Shark Bay, Australia *29*
succession 70
survival
  of the fittest 1, 2, 89–90, *89*
  short-term vs. long-term pressures 3
symbiosis 52

Telphusa, blood concentration *7*
termites, suicide committed by 3
themes, fundamental (principles) 62, *62*, *63*
tissue fluid formation *5*
tissue systems 68
topics 65, *65*
tree shrew *34*
Trichosomoides crassicauda 48
triploblastic animals 68
Tristan da Cunha 37, *37*

viruses, mutualism in 53

Wallace, Alfred Russell 39
white woodlouse (*Platyarthrus hoffmanseggii*) 54

Yucca, pollination of 54